THE DISPERSAL CENTRES OF TERRESTRIAL VERTEBRATES IN THE NEOTROPICAL REALM

BIOGEOGRAPHICA

Editor-in-Chief

J. SCHMITHÜSEN

Editorial Board

VOLUME II

DR. W. JUNK B.V., PUBLISHERS, THE HAGUE 1973

THE DISPERSAL CENTRES OF TERRESTRIAL VERTEBRATES IN THE NEOTROPICAL REALM

A study in the evolution of the Neotropical biota and its native landscapes

by

PAUL MÜLLER

DR. W. JUNK B.V., PUBLISHERS, THE HAGUE 1973

ISBN 90 6193 203 3

Dissertation for the degree of venia legendi in biogeography of the Philosophical
Faculty of the University of the Saarland, Saarbrücken (1970)

CONTENTS

I. INTRODUCTION

Although many scientists have studied the plants and animals of South and Central America we are still far from understanding their biogeography. The reason for this is only partly that: 'for population biologists who live and work in the temperate zone studies in tropical areas are expensive and difficult' (MACARTHUR & CONNELL, 1970, p. 45). A more profound reason is the extraordinary richness in plants and animals of the individual parts of this continent. Thus RICHARDS stated that: 'The most important single characteristic of the Tropical Rain Forest is its astonishing wealth of species' (1964, p. 229). HUMBOLDT, in 1859, confessed that the amazing diversity of the various biomes made it simply impossible to consider what these consisted of in detail.

Despite this there has been no lack of proposals for dividing the rich diversity of the South and Central American biota into zoogeographical regions and provinces (SCLATER 1858, WALLACE 1876, SCLATER & SCLATER 1899, MELLO-LEITÃO 1936, 1939, 1942, 1947, CABRERA & YEPES 1940, 1947, LANE 1943, RINGUELET 1955, 1961, HERSHKOVITZ 1958, 1969, RAPOPORT 1968, FITTKAU 1969, GERY 1969, KUSCHEL 1969, NOODT 1969, etc.). These attempts at subdivision are based on the geographical ranges of genera and families. The 'zoogeographical units' so produced are mostly only valid for one class of animals such as birds or fishes; they have no general significance (DARLINGTON 1957, DE LATTIN 1957). The anthropomorphic concept of the existence of sharply delimited faunal provinces—the regional concept—is put in doubt by the sheer complexity of the geographical ranges of species.

Even within the smallest areas there is often a complex interpenetration of lowland rain forest, montane forest, savanna and paramos. (This is shown for example by the distribution of vegetation of the Sierra de Santa Marta in Colombia.) As a result the species adapted to such areas show such a confusion of different geographical ranges that any attempt to classify these into types would be foolhardy.*

* The regional concept stands or falls with the demands that are made upon it. In the last analysis its problems are due to the fact that individual authors have made a historical concept of it (DARLING-TON, 1957). It is bad methodology to assume that the centre of range of a group or its centre of greatest diversity need be its centre of origin, but many authors have made this assumption (WALTER, 1954). Thus a North American species of a genus whose centre of species diversity is in the 'Neotropical Realm' is assumed to be of Neotropical origin. The Cracidae were considered by SCLATER (1858) and WALLACE (1876) as a Neotropical family. But later authors have described a series of fossil North American cracids, ranging in age from Eocene to Pleistocene (LUCAS 1900, WETMORE 1923, 1933, 1951, 1956, CHANEY 1947, BRODKORB 1954, TORDOFF & MACDONALD 1957). In consequence some authors now see it as a family of Nearctic origin (MAYR 1964, VUILLEUMIER 1965). In 1964 BRODKORB was nevertheless able to point to the existence of cracids in the Miocene of Argentina.

The analysis of dispersal centres offers a more satisfactory approach. The importance of such centres was already recognised by DE LATTIN for the Holarctic realm (1957, 1967). As explained under 'Methods', the analysis of such centres begins from a comparative study of animal and plant distribution. It differs from the regional concept in being based entirely on the geographical ranges of species and subspecies representing real systematic units (KOSSWIG 1959, MAYR 1967). All partial revisions yet carried out have confirmed the geographical position, evolutionary significance and generalisability of the centres analysed by DE LATTIN (GROSS 1961, 1962, WAGENER 1961, WARNECKE 1961, ANT 1964, ROESLER 1965, KEPCKA 1969, MÜLLER 1969, 1972, MÜLLER & SCHNEIDER 1969, WEIGEL 1969).

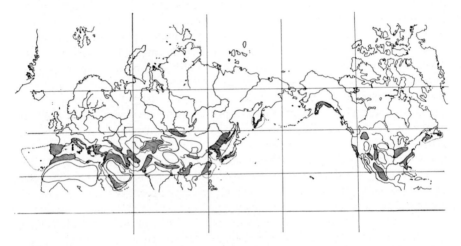

Fig. 1. Arboreal (= crosshatched) and eremial dispersal centres in the Holarctic realm (after DE LATTIN 1957).

The results got by analysing dispersal centres are not only important for biogeography. They also help us to understand how species arose, and throw light on earth and climatic history and present-day landscape relationships. Dispersal centres are as important for the study of evolution as for geography. For example, there is a dispersal centre in the Argentinian pampas, which is inhabited by endemic species adapted to an unforested biotope. This argues against the view that the pampas was forest land before the arrival of man (MÜLLER 1972).

In this work, therefore, I shall attempt to establish the existence of Neotropical dispersal centres. In this way I hope to contribute to the correct geographical subdivision of Central and South America and also to throw light on how the Neotropical biota arose.

II. METHODS

Dispersal centres can be worked out by plotting the breeding ranges of species and subspecies on a map of the region under investigation (DE LATTIN 1957). The individual ranges overlap in 'areas of congruence' or 'nuclear areas' ('Arealkerne' of REINIG 1937, 1950). It is our first task to ascertain regions where an unusually large number of ranges overlap, for these are what we call dispersal centres.

I do not assume at the outset that dispersal centres represent 'centres where faunas and floras were preserved during regressive phases'* (DE LATTIN 1957, p. 402), nor even that their origin is connected with the Pleistocene. I also do not wish to imply, *a priori*, that dispersal centres of particular taxa represent their centres of origin. To avoid misunderstanding, the two ideas must be kept quite separate.

Every species possesses, or used to possess, at least one dispersal centre that was its centre of origin. During the evolution of a taxon, however, the centre of origin and the centre of dispersal can become widely separated from each other (MÜLLER 1972).

Species and subspecies which resemble each other in their geographical distribution, and which can be ascribed to a single dispersal centre are the faunal elements of that centre in the sense of DE LATTIN (1957). The individual faunal elements can be classified according to how they relate to dispersal centres. Thus monotypic species, which can be related to only a single dispersal centre, are monocentric. Species related to several different dispersal centres are polycentric. In many of the cases which I have investigated polytypic species have a polycentric range made up of a number of monocentric subspecies ranges. In these cases the dispersal centre of a subspecies is in general the centre of origin of the population differentiated in it.

The analysis of dispersal centres presupposes that the species or subspecies ranges plotted on the map of a region are relatively small compared with the region itself. It also assumes that the limits of range are known with certainty and that the validity of the species or subspecies is not in dispute. These three conditions can be satisfied only among vertebrates and Lepidoptera even in regions like Europe and North America which are very well known from the systematic and distributional points of view.

What is more, the limits of the ranges of species and subspecies can often only be worked out with greater or lesser probability unless they happen to be stabilised by natural barriers such as water or mountains. As concerns the

* 'Erhaltungszentren von Faunen und Floren während regressiver Phasen'.

ranges of subspecies, working out the limits is made easier by the fact that populations differentiated only to a subspecific degree must have allopatric distributions for reasons of evolutionary genetics (MAYR, LINSLEY & USINGER 1953, MAYR 1967, MÜLLER 1972). Thus if we find subspecies A on the east bank of a river and subspecies B on the west bank, it is very likely that the river represents the boundary between the ranges of subspecies A and B. Species on the other hand, can be distributed sympatrically, and we could only show that the river was the boundary between two species on the basis of positive and negative evidence from both banks.

The validity of species and subspecies in the Neotropical realm has been best established among vertebrates, though of course with great differences between the taxa. And even among vertebrates many questions remain open. Birds have been well studied, but even among them 16 new Neotropical species were described between 1956 and 1965 (MAYR, 1971). The present work is therefore based mainly on terrestrial vertebrates—amphibia, reptiles, birds and mammals —which I have also studied intensively from the systematic point of view (MÜLLER 1966, 1967, 1968, 1969, 1970, 1971, 1972, 1973). In many species of reptiles and amphibia the basic information needed for distributional studies is lacking. Moreover, because of name changes it is impossible in these groups to compare old expedition reports with new ones without looking at voucher specimens and type specimens. The reptiles and amphibia are much worse in this respect than the 2965 species of birds exclusive to South America (MEYER DE SCHAUENSEE 1966). For this reason I first evaluated the systematic revisions and faunal lists cited in the Zoological Record since 1940 and only afterwards critically considered the older expedition reports.

Another difficulty was that, in many locality lists, there was no information about the gonadial development of birds. For this investigation only the breeding range of a species or subspecies is relevant, and consequently I have used such localities only when I could prove that they belonged to the breeding range of the species in question.

As well as examining the available literature I have been able to look through much unpublished material in various institutes and museums. When, because of lack of time, I was unable to work through material in an institute I have been able to have comparison material sent to me.

For some of the most interesting areas, especially Central Brazil and Amazonia, there were no locality data. In 1964, 1965, 1967 and 1969 I therefore undertook journeys of several months' duration especially to examine the vertebrate fauna of the South American islands, of the dry areas of Brazil and the 'islands' of campo within the Amazonian rain forest.

I should like to thank here all the institutions, friends, colleagues and teachers without whose help and advice this work would not have been possible. I have been helped with comparative material, information about localities and rare literature by the following gentlemen: Dr. R. ARLÉ, Museum Goeldi, Belem; W. BOKERMAN, Jardin Zoologico, São Paulo; Prof. D. DAREVSKI, Zoological Institute, University of Leningrad; Dr. J. EISELT, Naturhistorisches Museum,

4

Wien; Dr. J. HAFFER, Field Research Laboratory of Mobil Research and Development Corporation, Dallas, Texas; Prof. J. HAUSER, Faculdade de Filosofia, Ciencias e Letras, São Leopoldo, Rio Grande do Sul; Prof. W. HELLMICH, Sammlungen des Bayerischen Staates, München; Dr. A. R. HOGE, Instituto Butantan, São Paulo; Dr. M. S. HOOGMOED, Rijksmuseum van Natuurlijke Historie, Leiden; Dr. K. KLEMMER, Senckenberg Museum, Frankfurt-am-Main; the late Prof. E. MARCUS, Faculdade de Filosofia, Ciencias e Letras, São Paulo; Prof. E. MAYR, Museum of Comparative Zoology, Harvard University; Prof. R. MERTENS, Senckenberg Museum, Frankfurt-am-Main; Dr. G. PETERS, Zoologisches Museum der Humboldt Universität, Berlin; Prof. J. A. PETERS, Smithsonian Institution, Washington; Prof. P. SAWAYA, Universidade de São Paulo; Prof. B. SEHNEM, Faculdade de Filosofia, Ciencias e Letras de São Leopoldo, Rio Grande do Sul; Dr. H. SICK, Museum National, Rio de Janeiro; Dr. J. STEINBACHER, Senckenberg Museum, Frankfurt-am-Main; Dr. P. VANZOLINI, Departamento de Zoologiz, São Paulo; Dr. F. VUILLEUMIER, Station de Biologie marine, Roscoff, Prof. R. WEYL, Universität Gießen and Prof. E. WILLIAMS, Museum of Comparative Zoology, Harvard University.

As a guest I have spent long periods examining collections in many different museums. I should like especially to thank: the Naturhistorisches Museum in Vienna, the Museum Alexander Koenig in Bonn, the Museum do Homen do Sambaqui in Florianopolis, Brazil, the Museum Paulista and the Instituto Butantan in São Paulo, the Museum Goeldi in Belem, the Museum of the Faculdade de Filosofia, Ciencias e Letras in São Leopoldo and the Museum of the Instituto Nacional de Pesquisas da Amazonia (I.N.P.A.) in Manaus. I am also deeply grateful to Prof. C. KOSSWIG, Hamburg, with whom I have discussed problems of evolutionary genetics and Dr. W. F. REINIG, Hardt, for discussion of terminology.

Above all I should like to thank my teachers the late Prof. G. DE LATTIN and Prof. J. SCHMITHÜSEN. My own results, as presented in this work, are a synthesis of what I have learnt from them with what I have gleaned myself through a particular study of neotropical vertebrates and neotropical landscapes.

III. THE LIMITS OF THE AREA UNDER INVESTIGATION

It is necessary to define what the Neotropical realm means before its distribution centres can be investigated.

Many authors have divided the Americas into faunistic regions. The most important are SCLATER (1858), WALLACE (1876), SCLATER & SCLATER (1899), MELLO-LEITÃO (1939, 1942, 1946, 1947), CABRERA & YEPES (1940, 1947), LANE (1943), SIMPSON (1950, 1965), RINGUELET (1955, 1961), DARLINGTON (1957), MAYR (1964), DE LATTIN (1967), RAPOPORT (1968), HERSHKOVITZ (1969). Criticism of their various attempts has recently been concentrated on how the Nearctic and Neotropical realms should be divided from each other.

Central America is put by most authors into the Neotropical realm. Others, however, have considered it as a transition zone between the Neotropical realm and Nearctic region, though with a majority of Neotropical taxa (MAYR 1964, SIMPSON 1965, HERSHKOVITZ 1969, HOWELL 1969). Still others have made it an independent region alongside the other two (MERTENS 1952, KRAUS 1955, 1960, 1964, SAVAGE 1966). SAVAGE stated (1966, p. 719): 'Genera with South American distributions are poorly represented in Central America north of Panama and make up only 14% of the fauna north of Costa Rica. Under these circumstances the tropical Middle American assemblage is regarded as a distinctive unit, the Mesoamerican Herpetofauna, equivalent in rank to the Nearctic and Neotropical units.' (fig. 2.)

Part of the reason why individual authors have come to such different conclusions is the different powers of movement in the various groups they have worked on. A distinct Central American area is not clearly revealed by studying such mobile taxa as the birds or mammals (SIMPSON 1940, 1950, 1965, 1966, MAYR 1964, HOWELL 1969). On the other hand less mobile groups such as chilopods, diplopods, reptiles and amphibians, show it with more clarity.

Regardless of their views on the regional placing of the Central American fauna, all authors agree that in the lowland rain forests of Central America, the proportion of species of South American origin is astonishingly high. MERTENS (1952, p. 11)* said: 'It is characteristic that this Central American element in El Salvador is distributed not only in the lowlands, but also in the mountains, and this is probably true also for other parts of Central America. As against this, the purely Neotropical element is subordinate in the highlands, but *predominates* in lowland areas. This confirms my supposition that the Neo-

* 'Bezeichnenderweise ist dieses zentralamerikanische Element in El Salvador (und wohl auch in anderen Gebieten Mittelamerikas) nicht nur in den Tiefländern, sondern vor allem auch im Gebirge verbreitet, während das rein neotropische Element in den höhen Lagen zurücktritt und in der Ebene *vorherrscht*, entsprechend unserer Vorstellung von seiner ziemlich späten Einwanderung.'

6

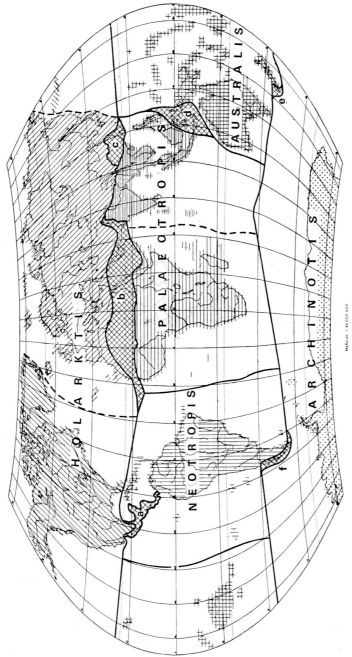

Fig. 2. Animal realms (from MÜLLER), Crosshatched areas a, b, c, d, e, f are transition zones.

7

tropical element entered the area fairly recently.' KRAUS (1960, p. 518)** like-wise wrote that: 'Endemic Central American faunal elements, particularly those

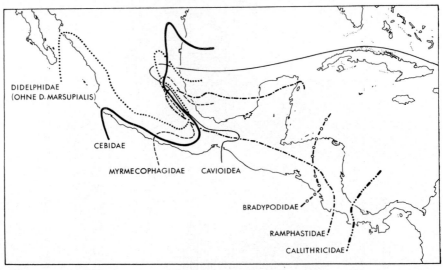

Fig. 3. The northern limits in Central America of families of South American origin. Note the barrier-zone in Mexico. The limit of the Didelphidae does not apply to *Didelphis marsupialis.*

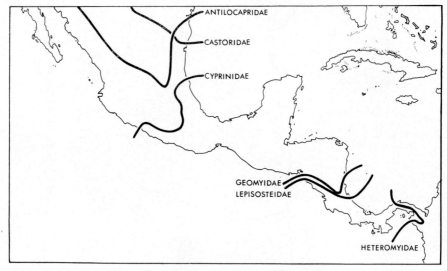

Fig. 4. The southern limits of North American families in Central America.

** 'Endemisch—zentralamerikanische Faunenelemente, dies gilt vor allem wieder für die Diplopo-den als extrem feine Indikatoren, besiedeln fast alle Lebensräume, von der heißen Ebene bis zu den Nebelwäldern der Vulkane und der Cordilleren. Es ist auffällig, wie sehr andere Elemente, die wir für junge Einwanderer halten, auf die heißen, tiefer gelegenen Zonen beschränkt sind.'

fine indicators the diplopods, live in almost all habitats, from the hot plain to the cloud forests of the volcanoes and the Cordillera. It is striking that other elements, which I take to be more recent immigrants, are restricted to the hotter, lower zones of altitude.'

In figs. 3 and 4 I have plotted on maps the northern limits of South American families of vertebrates and the southern limits of North American families. Two things emerge:

1. A large number of South American families have their northern limits in Central America, compared with the number of North American families that have their southern limits in this region.

2. There is a barrier-zone in Central America where a particularly large number of South American families have their northern limits.

This barrier-zone is correlated with the northern boundary of the Central American lowland rain forests and with the 1500 m. contour of the Sierra Madre in Mexico. BAKER (1963) and STUART (1966) have recorded other ecologically determined zones of the same sort for individual species, but these have no general significance.

The Neotropical realm in this work will be taken as the region lying south of this barrier-zone, where so many South American families have their northern limits. The northern boundary of the Neotropical realm is therefore correlated with the southern limit of the Mexican arboreal centre which DE LATTIN (1957) regarded as the southernmost dispersal centre of the Nearctic region. Starting from the northern limit of the Neotropical realm I shall analyse the dispersal centres southwards to Patagonia.

I shall not consider the West Indian Islands, including the Bahamas, although a high percentage of their recent and fossil fauna is of South American origin (BARBOUR 1914, 1930, 1937, GORMAN & ATKINS 1969, WILLIAMS & KOOPMAN 1951, 1952, SIMPSON 1956, KOOPMAN 1958, BOND 1960). The individual islands have been isolated from each other for so long (STANLEY 1970, VANDEL 1972) and have such a large number of endemics, that each one is an independent dispersal centre by definition. The most important works on this group of islands demonstrate this fact convincingly (FOWLER 1918, SCHARFF 1922, DANFORTH 1925, DAVIS 1926, DUNN 1926, COCHRAN 1928, 1941, GRANT 1932, DARLINGTON 1938, MERTENS 1938, 1939, MYERS 1938, 1950, SCHUCHERT 1935, ALAYO 1951, SMITH 1954, UNDERWOOD 1954, COOPER 1958, RIVAS 1958, GRIFFITHS 1959, SCHWARTZ 1960, 1964, 1965, 1966, 1967, SCHWARTZ & MCCOY 1970, COLLETTE 1961, GANS & ALEXANDER 1962, LAZELL & WILLIAMS 1962, LAZELL 1964, ETHERIDGE 1965, THOMAS 1966, 1968, GORMAN 1969).

A few of these Antillean centres represent true centres of origin e.g. of *Eleutherodactylus*, *Sphaerodactylus*, Todidae, though they were open to many outside influences, particularly during the Pleistocene (cf. KOOPMAN 1958).

IV. THE DISPERSAL CENTRES OF TERRESTRIAL VERTE-BRATES IN THE NEOTROPICAL REALM

A. THE GEOGRAPHIC POSITION AND FAUNAL ELEMENTS OF THE NEO-TROPICAL DISPERSAL CENTRES

1. The Central American rain-forest centre

The position of the centre (fig. 5, no. 1) can be defined by the overlap of the ranges of *Hyla ebraccata*, *Hyla microcephala* and *Heliconius hortense* (fig. 6).

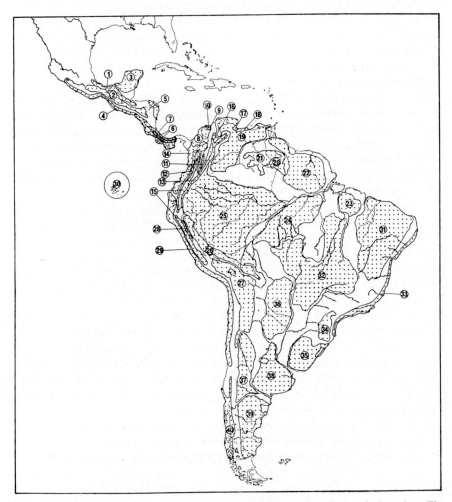

Fig. 5. The dispersal centres of terrestrial vertebrates in the Neotropical realm. The numbers indicating centres correspond to the numeration in the text.

It includes south-eastern Mexico, except for the dry region of Yucatan, and also northern Guatemala and central Honduras.

The limits of range of the faunal elements of the centre are determined, in most of the cases that have been studied, by three factors, i.e.:

a. the course of the 1500 m. contour,
b. the occurrence of pine forests in Honduras and Nicaragua (*Pinus caribaea* etc. cf. MIROV 1967),
c. the presence of unforested areas.

The limits of the Central American dispersal centre are therefore different from the limits of the Veracruz Province of SAVAGE (1966).

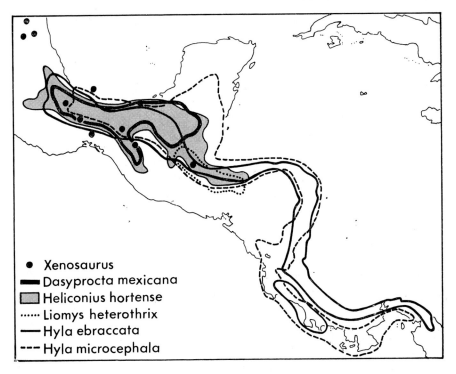

● Xenosaurus
▬ Dasyprocta mexicana
▦ Heliconius hortense
······ Liomys heterothrix
— Hyla ebraccata
--- Hyla microcephala

Fig. 6. The ranges of faunal elements of the Central American rain-forest centre. *Hyla ebrac-cata* and *Hyla microcephala* are polycentric species which occur in the Costa Rican centre as well as in the Central American rain-forest centre.

Of the 417 species of birds that occur in the centre, 59 species and 111 subspecies can be considered as faunal elements of the centre. (This statement is based on data from CHAPMAN 1923, CORY, HELLMAYR and CONOVER 1918–1949, PETERS 1931–1962, GRISCOM 1932, 1935, WETMORE 1943, 1944, 1959, 1963, SKUTCH 1944, 1954, 1958, 1960, 1964, 1967, ZIMMER 1945, GRISCOM 1932, 1935 LOWERY & DALQUEST 1951, EISENMANN 1955, 1962, HOWELL 1957, 1969, SLUD 1957, 1964, MONROE 1968.) The populations differentiated at a subspecies level in the centre belong to polycentric species. The 59 populations differentiated as species are monocentric.

The factors already mentioned as defining the ranges of most faunal elements have not been important for some elements. Thus some faunal elements of the Central American rain-forest centre extend into the montane forest zone by way of the valley of the river Chiapa that flows far to the south, e.g. *Dasyprocta montana*. Some species have also succeeded in going round the Sierra Madre to the north by way of the plain of Tehuantepec. On the western slope of the Pacific Coastal Range these species penetrate as far as El Salvador, where they occur in forest areas that are now isolated (*Eleutherodactylus rugulosus* etc.). Judging by the localities of occurrence that I know of, these species are always restricted to lowland rain forest. It is therefore likely that this migration route is no longer open (cf. the vegetation of El Salvador according to LAUER 1954, 1960).

Most of the subspecifically differentiated bird populations that can be regarded as faunal elements of this centre are most closely related to populations in the Costa Rican centre (82 out of the 111 cases = 74%). 22 populations distinct at the subspecies level have their closest relatives in the Central American montane forest centre and 7 on the Yucatan peninsula. The Yucatan subspecies that are related to populations in the Central American rain-forest centre are absent, nevertheless, in the dry area of the peninsula (cf. Yucatan centre).

Polycentric species which occur in the Central American rain-forest centre, but do not have subspecifically differentiated populations so far as known, are contradictory in distribution. In most cases, however, southern relationships and origin are indicated (Costa Rican centre).

The ranges of eurytopic species are not interrupted by the unforested areas of Nicaragua. These species occur along the coast, partly also on the banks of rivers and in places even in stands of *Pinus caribaea*. (*Hyla ebraccata* and *Hyla microcephala* have nevertheless not been found in the rain forests of Matagalpa). How far the ranges of these species have been changed by human influence can only be worked out by careful studies in the areas concerned. In this connection it is interesting to compare the ranges of *Pipra mentalis*, *Manacus candei* and *Columba nigrirostris* in HAFFER 1967 and *Coniophanes bipunctatus* in MYERS 1969.

The ranges of the herpetofauna of the centre are not so well established as those of the bird fauna, which have been better studied. Available data indicate that the reptiles and amphibians of the fauna have a richness in endemics comparable with that of birds (STUART 1935, 1950, 1951, 1963, 1964, 1966, SMITH & TAYLOR 1945, 1948, 1950, MERTENS 1952, DUELLMAN 1960, 1963, 1966, BRAME & WAKE 1963, NEILL 1965, ECHTERNACHT 1968, 1970, 1971).

Among the amphibia, even within the family Leptodactylidae, there are 23 species of *Eleutherodactylus* which must be regarded as faunal elements of the centre i.e. *Eleutherodactylus alfredi, E. anzuetoi, E. bocourti, E. brocchi, E. decoratus, E. dorsoconcolor, E. dunni, E. greggi, E. laevissimus, E. loki, E. merendonensis, E. milesi, E. polyptychus, E. ranoides, E. rhodopis, E. sammartinensis, E. rugulosus, E. sandersoni, E. spatulatus, E. stadelmani, E. stantoni, E. venustus, E. xucanebi*.

Consideration of the localities where they have been found shows that 18 species of snakes are also faunal elements of this centre i.e. *Typhlops basimaculatus, Leptotyphlops phenops phenops, Scaphiodontophis carpicinctus, S. cyclurus, S. nothus, Adelphicos quadrivirgatus quadrivirgatus, Dendrophidion vinitor, Dipsas dimidiatus, Drymobius chloroticus, Elaphe flavirufa flavirufa, E. triaspis, Lampropeltis triangulum polyzona, Pliocercus aequalis, Tantilla moestra, T. phrenitica, Tropidodipsas sartori sartori, Coniophanes quinquevittatus, Bothrops sphenophrys.* (The nomenclature is that of SMITH & TAYLOR 1945.)

The connections with the Costa Rican centre are also extremely close in the herpetofauna—thus *Bothrops nigroviridis nigroviridis* occurs in the Costa Rican centre while *Bothrops nigroviridis aurifer* is found in the Central American rain-forest centre. One sign of this close affinity is that at least 34 polycentric but monotypic species are endemics of both centres.

Nonetheless the Central American rain-forest centre lacks a number of genera which are represented by species in the Costa Rican centre. Such include: *Caecilia, Protopipa, Chiasmocleis, Relictivomer, Gastrotheca, Pleurodema, Cerathyla, Geochelone, Morunasaurus, Enyalioides, Echinosaura, Scolecosaurus, Amphisbaena, Anomalepis, Trachyboa, Atractus, Diaphorolepis, Lygophis, Phimophis, Pseudoboa, Siphlophis, Tripanurgos, Phyllobates, Prostherapis, Teratohyla, Atelopus, Glossostoma, Polychrus, Leposoma, Ptychoglossus, Neusticurus, Anadia, Helminthophis, Liotyphlops, Epicrates, Helicops, Leimadophis, Nothopsis, Dendrobates, Centrolenella, Corallus, Chironius, Erythrolamprus, Pseustes, Rhinobothryum, Lachesis.*

On the other hand at least 19 genera widely distributed in South America have reached the Central American rain-forest centre. In most cases they are represented by clearly differentiated species or subspecies. This group includes: *Leptodactylus, Engystomops, Cochranella, Ameiva, Gonatodes, Gymnophthalmus, Typhlops, Clelia, Dendrophidion, Dipsas, Dryadophis, Imantodes, Oxyrhopus, Xenodon, Caiman, Micrurus, Bothrops, Spilotes, Mabuya.*

Eighteen of the genera of reptiles and amphibia which are widely distributed in the Nearctic region also occur in Central America. But of these only *Storeria* and *Rana* are represented in the Central American rain-forest centre. In SAVAGE's opinion (1966) *Rana* and *Storeria* both entered Central America from North America before the Pleistocene. All the other genera in question (*Crotalus, Agkistrodon, Eumeces, Gerrhonotus* etc.) occur either in the unforested biotopes or in the montane forests and *Pinus* forests of Central America, or in a few cases also in the paramos of Costa Rica.

Twenty-three other genera that have species in the Central American rain-forest centre, cannot certainly be attributed to the Nearctic region or to the Neotropical realm. Such cases are either world-wide genera (*Bufo, Hyla*) or Central American endemics (*Plectrohyla, Ptychohyla*). SAVAGE's assertion (1966) that Central America was the centre of origin of these endemics is based purely on the absence of fossils of these taxa in the Nearctic region and South America.

2. The Central American montane forest centre

The position of this centre is defined by the ranges of *Ptychohyla euthysanota* and *Hyla uranochroa* (fig. 7, fig. 5, no. 2). The ranges of these two species show

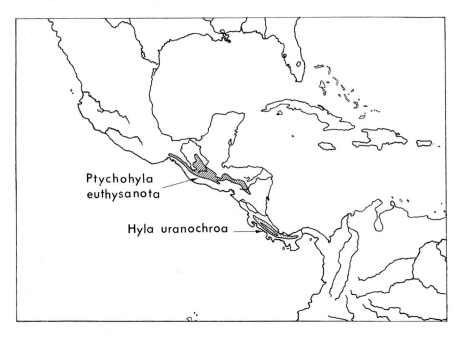

Fig. 7. The ranges of *Ptychohyla euthysanota* and *Hyla uranochroa*. These two elements of the Central American montane forest centre show the positions of the Talamanca montane forest subcentre and of the Guatemalan montane forest subcentre.

that two subcentres can be distinguished. One of these is situated in the montane forests of northern Panama and northern Costa Rica; this is the **Talamanca montane forest subcentre** and corresponds to the range of *Hyla uranochroa* in fig. 7. The other lies in the montane forests which stretch from Nicaragua into southern Mexico; this is the **Guatemalan montane forest subcentre** and corresponds to the range of *Ptychohyla euthysanota* in fig. 7.

The relationship between the two subcentres is even closer than the relationship between the Central American and Costa Rican rain-forest centres. Thus the following species of snake occur in both subcentres: *Amastridium sapperi, Chersodromus liebmanni, Geophis nasalis, Tropidodipsas fischeri, Thamnophis sumichrasti fulvus, Bothrops undulatus* and *B. godmanni*.

The faunal elements of the centre are closely restricted to the montane forests i.e. above 1500 m. This is shown by the ecological studies that have been made on the amphibia and birds of the centre (MCCOY & WALKER 1966, ANDRLE 1967, SKUTCH 1967, VIAL 1968, MYERS 1969).

Compared with the herpetofauna, the bird fauna of the centre has been well investigated. (CABANIS 1860–1862, 1869, FRANTZIUS 1869, BANGS 1909, RENDAHL 1918–1920, PETERS 1931–1962, ALDRICH & BOLE 1937, DICKEY & VAN ROSSEM 1938, GRISCOM 1932, GOODWIN 1946, BLAKE 1953, 1958, EISENMANN 1955, WAGNER 1961, DEIGNAN 1961, BAKER 1963, WEBB & BAKER 1962, SLUD 1964, SKUTCH 1967).

Forty-four species of birds can be considered as faunal elements of the Talamanca montane forest subcentre (compiled from SLUD 1964). Polycentric species are represented in 74 cases by monocentric subspecies. Of these, 14 correspond to the southern limit of their species and 37 to the northern limit. Twenty-three of the populations differentiated at subspecies level have their closest relatives in the Guatemalan montane forest subcentre and in the Colombian montane forest centre.

The Guatemalan montane forest subcentre corresponds to the 'Guatemalan Highland Province' of SAVAGE (1966). Forty-eight endemic species and 58 endemic subspecies confirm the distinctness of this subcentre. Of the 58 endemic subspecies 37 have their closest relatives in the Talamanca montane forest subcentre and 21 in the Mexican centre of DE LATTIN (1957, 1967). For comparison the herpetofauna of the two subcentres shows the following distribution:

	Faunal elements of the Guatemala subcentre	Faunal elements of the Talamanca subcentre
Caudata	14	9
Anura	28	17
Sauria	14	6
Serpentes	16	13
	72	45

In terms of origin, 7 of the Caudata of the Guatemalan subcentre point to the Mexican centre and 7 to the Talamanca subcentre. The 9 faunal elements among the Caudata of the Talamanca subcentre have their nearest relatives in the Guatemalan subcentre (7 cases) or in the Colombian montane forest centre (2 cases). The same pattern seems to be confirmed by the other reptiles and the Anura. Here, however, the affinities with other centres have only been satisfactorily worked out in a few taxa.

The proportion of amphibians and reptiles with Nearctic relationships is higher than in the Central American rain-forest centre. This is also true for birds and mammals.

3. The Yucatan centre

This centre (fig. 5, no. 3) can be defined from the ranges of: *Campylorhynchus yucatanicus* (Aves, Troglodytidae), *Myiarchus yucatanensis* (Aves, Tyrannidae), *Agriocharis ocellata* (Aves, Meleagridae, cf. range in LEOPOLD 1959), *Peromyscus yucatanicus* (Mammalia, Rodentia), *Eleutherodactylus laticeps* (Amphibia, Leptodactylidae), *Enyaliosaurus defensor* (Reptilia, Iguanidae) and *Bothrops yucatannicus* (Reptilia, Crotalidae) (fig. 8). These species are ecologically very closely adapted to the dry savanna and thorn-bush savanna of Yucatan (for the birds cf. HOWELL 1969 and EISENMANN 1955).

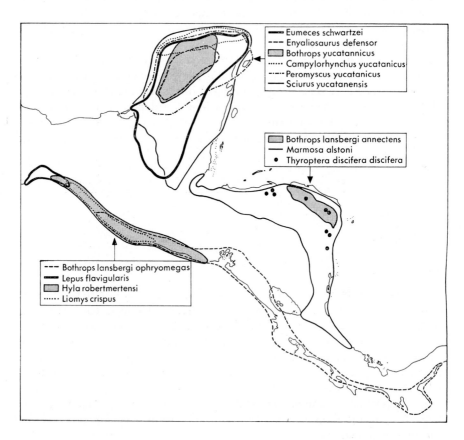

Fig. 8. Ranges of faunal elements of: the Yucatan subcentre—*Eumeces schwartzei, Enyaliosaurus defensor, Bothrops yucatannicus, Campylorhynchus yucatanicus, Peromyscus yucatanicus, Sciurus yucatenensis*: of the Coco centre—*Bothrops lansbergi annectens, Marmosa alstoni, Thyroptera discifera discifera*: and of the Central American Pacific centre—*Bothrops lansbergi ophryomegas, Lepus flavigularis, Hyla robertmertensi, Liomys crispus*.

Besides such stenotopic faunal elements there are eurytopic ones which also occur in the humid savanna of Yucatan (*Eumeces schwartzei, Sciurus yucatanensis*). These are mainly not represented by distinct subspecies.

Those that are subspecifically differentiated, however, have one subspecies that indicates the humid savanna (e.g. *Cnemidophorus angusticeps petenensis*) while the other is an endemic of the dry savanna and thorn-bush savanna (cf. BEARGIE & MCCOY 1964, WRIGHT & LOWE 1968).

In addition to this a study of the distributions of species adapted to rain forest or to savanna shows that the area between Flores (Guatemala) and Xcalak (Mexico) is a zone of interpenetration between the Yucatan centre and the Central American rain-forest centre. Full species endemic to this zone of interpenetration are unknown. It is inhabited by eurytopic rain forest or savanna species. The Iguanid *Laemanctus serratus* is an example of this. *L. serratus serratus* lives in the Central American rain forest while a subspecies derived from this—*L. s. alticoronatus*—has successfully colonised the transition zone.

There are isolated breeding populations of indicator species of unforested habitats in the middle of the rain-forest biome of Guatemala and Mexico (north of the Rio Usumacinta). These, however, are only weakly differentiated e.g. *Crotalus durissus, Rhinophrynus dorsalis*. They represent species or subspecies which immigrated from the Yucatan centre. PAYNTER (1955) supposes that most of the bird species of this group only penetrated this area after the forests had been destroyed by man (see, for example, the range of *Colinus nigrogularis* in PAYNTER 1955). He says (p. 81): 'Because of the extensive cultivation of henequén in the vicinity of Mérida, and the presence of Indians who practice milpa agriculture in the outlying districts, the present area of habitats suitable for quail is vastly greater than at any time during the history of the Peninsula.'

The same relationships exist in the Colubrid genus *Coniophanes*. Five species occur in the Yucatan peninsula i.e. *C. bipunctatus, C. imperialis clavatus, C. meridanus, C. quinquevittatus* and *C. schmidti*, according to data in MCCOY (1969) and DUELLMAN (1965). An analysis of the dispersal centres, however, reveals only two of these as Yucatan faunal elements closely adapted to the dry areas i.e. *C. schmidti* and *C. meridanus*. *C. meridanus* nevertheless also occurs together with *Crotalus durissus tzabcan* in the isolated savanna 'islands' in the rain forest (KLAUBER 1952, *Bull. Zool. Soc.* San Diego 26: 71). The three other *Coniophanes* species should be reckoned as faunal elements of the Central American or of the Costa Rican rain-forest centres.

There are 107 known species of reptiles and amphibia within the area of the Yucatan Peninsula (i.e. in an area of 143,500 km^2) excluding the three marine turtles *Caretta caretta, Chelonia mydas* and *Eretmochelys imbricata*. A cautious analysis of these 107 species suggests that 46 should be considered as faunal elements of the Yucatan centre and 27 as faunal elements of the Central American rain-forest centre. Twelve species are widely distributed with a definite preference for an unforested biotope. All these twelve occur also in the Central American Pacific centre and have close affinities with the Nearctic region. This analysis is based on data in COLE & BARBOUR (1906), BAILEY (1928), FRITTS (1969),

HARTWEG (1934), GAIGE (1936), SCHMIDT (1936), SCHMIDT & ANDREWS (1936), ANDREWS (1937), SMITH (1938, 1947), STEJNEGER (1941), KLAUBER (1945), SMITH & TAYLOR (1945), PETERS (1953), STUART (1958), NEILL & ALLEN (1959, 1962), DOWLING (1960), BARRERA (1962), DUELLMAN (1963, 1965), MASLIN (1963), MCCOY (1963, 1968), WELLMAN (1963).

Of the 46 faunal elements of the Yucatan centre 31 have close relatives in the Central American Pacific centre. Two have their closest relatives in the Coco and Barranquilla centres and 13 in Nearctic distribution centres. The faunal elements of the Yucatan centre show a closer relationship to those of the Central American Pacific centre than to those of North or South American dispersal centres.

It is interesting that of the 47 species of snake in Yucatan, as many as 22 are faunal elements of the centre: *Typhlops microstomus, Dipsas brevifacies, D. sanniolus, Dryadophis melanolomus melanolomus, Imantodes splendidus splendidus, I. tenuissimus, Leptodeira yucatanensis yucatanensis, Leptophis mexicanus yucatanensis, Ninia sebae morleyi, Opheodrys mayae, Pliocercus andrewski, P. elapoides schmidti, Tantilla canula, T. cuniculator, Tropidodipsas fasciata, Thamnophis sumichrasti praeocularis, Coniophanes schmidti, C. meridanus, Micrurus affinis mayensis, Bothrops yucatannicus, Crotalus durissus tzabcan, Agkistrodon bilineatus taylori.*

4. The Central American Pacific centre

This centre (fig. 5, no. 4) can be defined on the ranges of *Hyla robertmertensi* (Anura, Hylidae), *Scaphiodontophis albonuchalis*, *Imantodes gemmistratus* (Reptilia, Colubridae) and *Bothrops lansbergi ophryomegas* (Reptilia, Crotalidae) (fig. 8).

I mention in passing that unlike HOGE (1966), I do not see *B. l. ophryomegas* as a separate species, but as a subspecies of *Bothrops lansbergi*. The reason for this is partly the great morphological resemblance between *B. l. ophryomegas* and *B. l. annectens*, which is a faunal element of the Coco Centre (cf. fig. 8); in addition this systematic placing is supported by the distribution of the other subspecies of *B. lansbergi*, as I shall discuss further elsewhere.

Of the 632 species of amphibia and reptiles known in Central America, at least 178 occur in the Central American Pacific centre. Fifty-nine of these must definitely be taken as faunal elements of the centre. This statement is based on data from FOWLER (1913), DUNN (1931, 1940, 1949), STUART (1935, 1950, 1951, 1954, 1958, 1963), BARDEN (1943), EVANS (1947), SMITH & TAYLOR (1945, 1948, 1950), TAYLOR (1951, 1952, 1954, 1955, 1956, 1958), MERTENS (1952), CARVALHO (1954), DUELLMAN (1956, 1960, 1963, 1965, 1966), SHREVE & GANS (1958), BRAME & WAKE (1963), LEGLER (1963), CAMPBELL & HOWELL (1965), LYNCH &

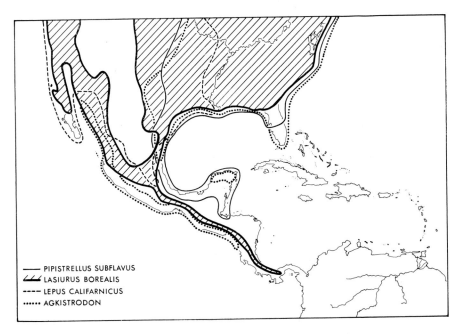

Fig. 9. The ranges of some Nearctic taxa in Central America. Note the preference of these taxa for the Central American montane forest centre, the Yucatan centre and the Central American Pacific centre.

FUGLER (1965), NEILL (1965), SEXTON & HEATHWOLE (1965), ZWEIFEL (1965), SAVAGE (1966), HARDY & MCDIARMID (1969). Among such faunal elements is a monotypic Boid genus (*Loxocemus bicolor*, cf. STIMPSON 1969) whose southern limit coincides with that of *Bothrops lansbergi ophryomegas* and whose northern limit coincides with that of *Hyla robertmertensi*. The Boid *Ungaliophis continentalis*, which is the only known species of its genus apart from one in the Costa Rican and Columbian Pacific centres, has a similar distribution, as also does the Crotalid *Agkistrodon bilineatus bilineatus*.

The closest relationships of the faunal elements of the Central American Pacific centre are with the Yucatan centre (cf. discussion of that centre and the distribution of *Agkistrodon* in fig. 9). Thus the monotypic species *Rhinophrynus dorsalis* occurs in both these centres, but is lacking in the Central American rain-forest centre. Some polytypic species like *Agkistrodon bilineatus* have one monocentric subspecies in the Central American Pacific centre (*A. b. bilineatus*) and another in the Yucatan Centre (*A. b. taylori*); see BURGER & ROBERTSON (1951, *Univ. Kansas Sci. Bull.* 34: 213 and 125: 1–2). These species are absent in the Central American rain-forest centre. Monocentric species of the Central American Pacific centre, such as *Enyaliosaurus quinquecarinata* or *Eumeces manague*, are replaced in the Yucatan centre by the species closest related to them (*Enyaliosaurus defensor*, *Eumeces schwartzei*).

No close relationship to the Central American rain-forest centre can be established. The relationship to the Coco centre is also rather slight, but this reflects the small total number of faunal elements which can be reckoned to that centre. Thus four birds, one mammal and two snakes of the Central American Pacific centre are replaced by conspecific subspecies in the Coco centre. A still unsettled question in this connection is whether *Crotalus durissus* occurs in the Coco centre. KLAUBER supposed that it did (1936, Occ. Pap. San Diego Soc. Nat. Hist., 1. Rattlesnakes 1, p. 4, 1956, p. 31–32) but I have found no specimens to support his view.

There are closer connections with the Barranquilla centre. Thus 14 birds, three snakes and four lizards of the Central American Pacific centre are replaced by conspecific subspecies in the Barranquilla centre. The relationships with the Gila and Mexican centres of the Nearctic region (cf. DE LATTIN 1957, 1967) is weaker than with the Yucatan centre, both for the birds and for the amphibia and reptiles.

A weak barrier-zone to the faunal elements of the Central American Pacific centre exists in the region of Tehuantepec. Thirty-three such faunal elements among the reptiles and amphibia have their northern limits in this place.

DUELLMAN (1960) worked on the amphibian fauna of 70 localities in the plain of Tehuantepec from the Pacific to the Atlantic side and established the presence of 36 species. These were: *Gymnopis mexicanus mexicanus*, *Bolitoglossa occidentalis*, *B. platydactyla*, *B. veracrucis*, *Rhinophrynus dorsalis*, *Bufo canaliferus*, *B. coccifer*, *B. marinus*, *B. marmoreus*, *B. valliceps*, *Eleutherodactylus alfredi*, *E. natator*, *E. rhodophis*, *E. rugulosus*, *Microbatrachus pygmaeus*, *Syrrhopus leprus*, *S. pipilans*, *Engystomops pustulosus*, *Leptodactylus labialis*, *L. melano-*

notus, Diaglena reticulata, Hyla baudini, H. ebraccata, H. loquax, H. micro-cephala martini, H. picta, H. robertmertensis, H. staufferi, Hylella sumichrasti, Phrynohyas modesta, P. spilomma, Phyllomedusa callidryas taylori, P. dacnicolor, Gastrophryne usta, Rana palmipes, R. pipiens.

The distribution and preferred biotopes of these species indicate that the ecological character of the plain has fluctuated greatly in the past (cf. Discussion). As to distribution, nine species are known on the Atlantic side of the plain only, six on the Pacific side only, two only on the plain proper (*B. veracrucis* and *H. sumichrasti*) and nineteen on the Pacific and Atlantic sides; while as to biotope, eleven are pure forest species, eighteen savanna species and seven distributed irregularly. This matter is treated further in the Discussion.

5. The Coco centre

The faunal elements of the centre show that it has the same limits and distribution as the pine savanna in Honduras and Nicaragua (fig. 8 and fig. 5, no. 5). Floristically it coincides with the south-eastern occurrence of *Pinus caribaea*.

There is a striking absence of endemic species, with the possible exception of *Marmosa alstoni* (cf. fig. 8). Seventeen populations of birds differentiated at subspecific level have their closest relatives in the Yucatan centre (11 cases), the Central American Pacific centre (4 cases) or in the Central American rain-forest centre (2 cases).

The name of the centre comes from the Rio Coco, which separates Honduras from Nicaragua. The centre is not as large as the Caribbean Province of SAVAGE (1966) which involves in addition the rain-forest biome of Honduras and Nicaragua. Only eurytopic species, like *Marmosa alstoni* which penetrates the rain forest in places, could serve as indicators of this Caribbean Province. The distribution of stenotopic species, on the other hand, disagrees with SAVAGE's concept e.g. *Bothrops lansbergi annectens*, *Thyroptera discifera discifera* (cf. fig. 8).

Moreover it is still not known in enough detail how far species like *Marmosa alstoni* represent followers of man, whose distribution indicates where rain forest has been changed by human influence into 'artificial steppe.' I have considered elsewhere the problem of the expansion of original savanna species into these artificial savannas created by man (MÜLLER 1970).

During the maximum advance of ice in the Pleistocene a peninsula formed on the Caribbean coast of the Coco centre and extended a long way towards Jamaica (FAIRBRIDGE 1962, FRAY & EWING 1963, DE LATTIN 1967). This allowed many species to spread into the Antilles (e.g. LANYON (1967) for the evolution of the Tyrannid *Myiarchus* in the West Indies).

The degree of relationship of certain faunal elements of the Coco centre with those of Cuba, Jamaica and Haiti depends partly on the mobility of the group in question. The degree of relationship gets gradually weaker in the series, birds, bats, other mammals, reptiles, salt-water sensitive amphibia (cf. COCHRAN 1928, DARLINGTON 1938, LAZELL 1964, ETHERIDGE 1965, LANYON 1967, GORMAN 1969). Some authors have postulated land connections between the Antilles and the Honduras-Nicaragua region, on the basis of the closeness of relationships among recent organisms (e.g. STUART, 1966). Before such attempts are made however, it is imperative to consider critically both the ecological valency of the 'indicator species' and the extent to which they can drift passively (PALMÉN 1944, MACARTHUR & WILSON 1967, WILSON & SIMBERLOFF 1969). The need for caution is emphasised by the fact that some taxa are extremely close, being differentiated only at subspecific level, while others lack relationship altogether.

6. The Costa Rican centre

The nucleus of this centre, in the sense of REINIG (1950), is the rain-forest biome north and south of the Sierra of Talamanca (fig. 5, no. 6). The northern limits of its faunal elements (figs. 10, 11) correspond to an obvious barrier-zone in the area of the Rio San Juan and Lake Nicaragua. A similar barrier-zone exists to the south-east, in the region of the Panama basin.

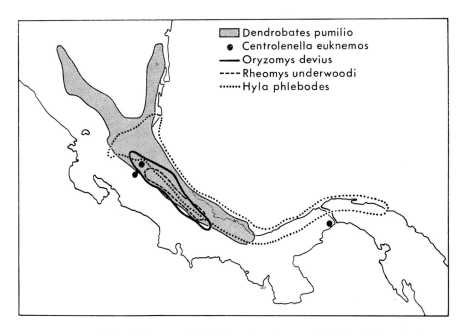

Fig. 10. Ranges of faunal elements of the Mosquito subcentre of the Costa Rican centre.

The clarity of the boundaries of the centre is very striking. This is true not only for the groups specially investigated here, but also true, for example, of the fishes. A distributional analysis of fresh-water fishes in Central America, excluding forms that have entered fresh water from the sea, shows that, among 18 families and 104 species of South American origin which occur (MILLER 1966, MYERS 1966) 12 families with 74 species in all are restricted to the Costa Rican centre i.e. are found only in Costa Rica and western Panama. This great richness of endemic fishes, whose distribution is controlled by quite other factors than the birds or reptiles, shows that this centre has had a different geological history from the other Central American centres.

The number of amphibia which can be taken as faunal elements of the centre is also astonishingly high. There are 35 faunal elements of the Costa Rican centre even within the family Leptodactylidae: *Eleutherodactylus altae, E.*

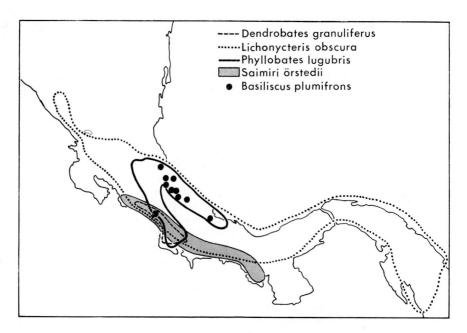

Fig. 11. Ranges of faunal elements of the Costa Rican centre. The ranges of *Saimiri oerstedii* and *Dendrobates granuliferus* indicate the position of the Chiriqui subcentre.

bransfordii, E. brederi, E. caryophyllaceus, E. cerasinus, E. costaricensis, E. crassidigitus, E. dubitus, E. fleischmanni, E. florulentus, E. gaigei, E. gollmeri, E. hylaeformis, E. lutosus, E. melanostictus, E. mimus, E. monnichorum, E. noblei, E. nubilus, E. pardalis, E. peraltae, E. pittieri, E. platyrhynchus, E. podiciferus, E. punctariola, E. rearki, E. ridens, E. rugosus, E. stejnegerianus, E. talamancae, E. taurus, E. underwoodi, E. vocator, Leptodactylus maculilabris, L. quadrivittatus.

The Costa Rican centre can be divided into two subcentres. North of the Sierra de Talamanca is the **Mosquito subcentre**, named after the Gulf of Mos-quitos which lies north of it. South of the Sierra de Talamanca is the **Chiriqui subcentre**, named after the Chiriqui province of RYAN (1963).

For rain-forest species which penetrate the montane forest zone, and to that extent are mobile, this secondary division naturally does not exist. Instances of this are *Dendrobates auratus, D. granuliferus, Lichonycteris obscura, Cerathyla panamensis* and other examples in SAVAGE (1968). But faunal elements which are strictly adapted to the lowland rain-forest biome do show this secondary subdivision of their ranges (cf. fig. 10, 11). The Mosquito subcentre is richer in species than the Chiriqui subcentre. This obviously reflects the differing histories of the Atlantic and Pacific rain forests (cf. Discussion).

There are isolated occurrences of Central American Pacific faunal elements in the peninsula of Azuero Santos, partly as subspecifically distinct populations. This indicates that there must have been, at least for a short time, an unforested migration route along the Pacific coast for species requiring an open landscape. Examples are: *Crypturellus cinnamomeus praepes, Emberizoides herbicola lucaris, Aimophila rufescens hypaethrus, Arremonops rufivirgata superciliosa, Amazilia boucardi, Crotalus durissus.*

Besides a close relationship to the Central American rain-forest centre (see under that centre), the faunal elements of the Costa Rican centre are closely related to the Colombian Pacific centre. This was already clearly recognised by CHAPMAN (1917) and was discussed by HAFFER (1967). There are 37 polycentric and polytypic species of birds with monocentric subspecies in the Costa Rican and Colombian Pacific centres. Examples are: *Columba subvinacea subvinacea* (Chiriqui subcentre) and *C. s. berlepschi* (Choco subcentre of the Colombian Pacific centre); *Tangara florida florida* (Mosquito subcentre) and *T. f. auriceps* (Choco subcentre). Besides these there are monotypic species which occur in both centres e.g. *Bothrops schlegelli, B. nasutus, B. asper, Cerathyla panamensis* (cf. MYERS 1966), *Eleutherodactylus moro* (cf. SAVAGE 1968, SAVAGE & HEYER 1967). In such species ascription to a dispersal centre has to be worked out case by case. Only fossils, or the sorting out of the phylogenetic connections of the taxon, give any useful indication.

7. The Talamanca paramo centre

This centre (fig. 5, no. 7) coincides with the distribution of the paramo of Costa Rica. Floristically it is closely related to the North Andean centre (WEBER, 1958, 1969). This connection is also very obvious from the faunal elements of the centre which in many cases have their closest related species or subspecies in the North Andean centre (fig. 5, no. 15).

Up till now only the bird fauna has been properly studied. However, of the 758 species of birds which occur in Costa Rica, with a total of 864 different forms placed in 440 genera (SLUD, 1964), only 21 can be considered as faunal elements of the Talamanco paramo centre. These are: *Zenaida macroura turtilla, Otus clarkii, Glaucidium jardinii costaricanum, Aegolius ridgwayi ridgwayi, Caprimulgus saturatus, Colibri thalassinus cabanidis, Lampornis hemileucus, Philodice bryantae, Selasphorus simoni, Philydor rufus panerythrus, Pachyramphus versicolor costaricensis, Myiodynastes hemichrysus, Contopus ochraceus, Acrochordopus zeledoni zeledoni, Cistothorus platensis lucidus, Tangara dowii, Chlorospingus zeledoni, Spodiornis rusticus barrilescensis, Melozone biarcuatum cabanisi, M. leucotis leucotis, Acanthidops bairdi*. Of these 21 faunal elements, only two belong to families of probably South American origin (MAYR 1964). These are *Philydor rufus panerythrus* of the Furnariidae and *Pachyramphus versicolor costaricensis* of the Cotingidae. All other populations are certainly of Nearctic or Central American origin. *Zenaida macroura* (= mourning dove) is an example of this, for the two subspecies closest related to *Z. m. turtilla* are *Z. m. carolinensis* and *Z. m. marginella* which occur in North America.

These results are quite opposite to those from the Costa Rican centre, whose faunal elements in great majority are of Neotropical origin. Thus of the 758 species of birds found in Costa Rica, 620 have their closest relatives in South or Central American centres.

Of the six species of bird which, according to SLUD (1964), occur only in Costa Rica, three are faunal elements of the Talamanca paramo centre i.e. *Selasphorus simoni, Chlorospingus zeledoni* and *Acanthidops bairdi*.

The subspecies *Melozone biarcuatum cabanisi* and *Melozone leucotis leucotis* are restricted to the paramos of Costa Rica. The other faunal elements extend as local populations into the northern parts of the mountains of Panama. Such populations, however, are not subspecifically distinct from the Costa Rican populations. This indicates that gene exchange must have taken place until a relatively short time ago, or perhaps still goes on.

The amphibians, reptiles and mammals now known from the centre have their closest relationships, like the birds, with Nearctic centres. Among reptiles it is possible to show an immigration route from the Mexican centre to the Talamanca paramo centre, by way of the pine-stands in the Central American highlands. See in this connection the ranges of *Gerrhonotus viridiflavus, G. moreleti* and *G. monticola* and of *Sceloporus formosus* and *S. malachiticus* in fig. 12.

Relationships with the North Andean centre are discussed under that centre.

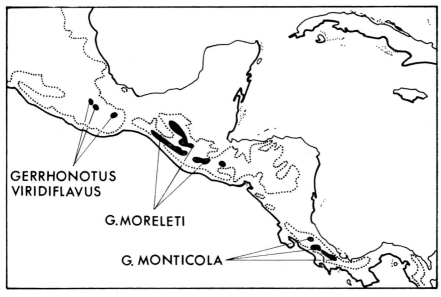

GERRHONOTUS
VIRIDIFLAVUS

G. MORELETI

G. MONTICOLA

········ 600–1500 m

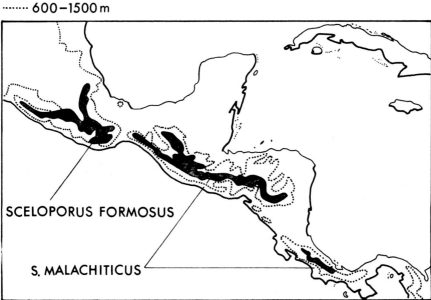

SCELOPORUS FORMOSUS

S. MALACHITICUS

Fig. 12. Ranges of species of *Gerrhonotus* (above) and *Sceloporus* (below). *Gerrhonotus monti-cola* is a faunal element of the Talamanca paramo centre whose closest relatives live in the Central American montane forest centre and in the Mexican centre. *Sceloporus malachiticus* is a polycentric faunal element found in the Central American montane forest centre and the Talamanca paramo centre. The species closest related to it is found in the Mexican centre. The populations of *Sceloporus* and *Gerrhonotus* in the Talamanca paramo centre correspond to the southern limits of these genera.

8. The Barranquilla centre

The position of the centre (fig. 5, no. 8) can be defined on the ranges of *Bothrops lansbergii lansbergii* (Reptilia, Crotalidae), *Micrurus dissoleucus nigrirostris* (Reptilia, Elapidae), *Sylvilagus brasiliensis sanctaemartae*, *S. floridanus superciliaris* (Lagomorpha), *Sciurus granadensis gerrardi* (Rodentia, Sciuridae), *Sanguinus oedipus oedipus* (Callithricidae), *Myiarchus ferox panamensis* (Aves, Tyrannidae), *Campylorhynchus griseus albicilius* (Aves, Troglodytidae), *Poecilurus candei candei* (Aves, Furnariidae) and *Aratinga pertinax griseipecta* (Aves, Psittacidae) (cf. fig. 13). It completely lacks endemic species (cf. the Coco centre).

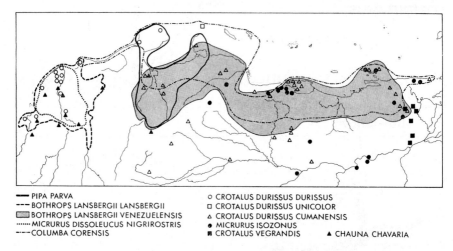

— PIPA PARVA
--- BOTHROPS LANSBERGII LANSBERGII
▓ BOTHROPS LANSBERGII VENEZUELENSIS
······ MICRURUS DISSOLEUCUS NIGRIROSTRIS
—··— COLUMBA CORENSIS

○ CROTALUS DURISSUS DURISSUS
□ CROTALUS DURISSUS UNICOLOR
△ CROTALUS DURISSUS CUMANENSIS
● MICRURUS ISOZONUS
▣ CROTALUS VEGRANDIS   ▲ CHAUNA CHAVARIA

Fig. 13. Faunal elements of: the Barranquilla centre (*Micrurus dissoleucus nigrirostris, Bothrops lansbergii*); and of the Caribbean centre (*Crotalus durissus cumanensis, Micrurus isozonus, Crotalus vegrandis, Bothrops lansbergii venezuelensis, Columba corensis, Pipa parva*). *Chauna chavaria* is a polycentric faunal element whose range indicates the close relationship between the Barranquilla and Caribbean centres.

Of the 23 subspecifically distinct bird populations, 18 have their nearest relatives in the Cauca, Magdalena, Coco, Central American Pacific and Caribbean centres. Five have their closest relatives in the Nechi and Choco subcentres of the Colombian Pacific centre (data from MEYER DE SCHAUENSEE, 1964). Similar relationships hold for the reptiles and mammals.

It is remarkable that the Barranquilla populations of *Crotalus durissus* extend as far as the Guajira peninsula, whereas the species is represented in the Maracaibo subcentre by the subspecies *C. durissus cumanensis* which is typical for the Venezuelan coast region. It should be noted in this connection that the Barranquilla populations are more closely related to the Central American nominate form than to *C. d. cumanensis*.

28

On Aruba, which is only 27 km. from the Paraguana peninsula, occurs the race *C. d. unicolor*, which is clearly distinct from both of the above subspecies (fig. 13). But on the island of Margarita, which is separated from the mainland by a sea only 30 m. deep, the subspecies *C. d. cumanensis* occurs once more.

However it would be highly premature to see the existence of *Crotalus durissus* on Aruba as indicating a Pleistocene land connection of this island. Equally premature is the assertion of VOOUS (1955) that the island was never connected with the mainland. This was based on the lack of Dendrocolaptidae, Furnariidae and Formicariidae which are families of South American origin (MAYR 1964). The lack of these typical forest-dwelling families of birds merely shows that no suitable ecological conditions exist for them on Aruba.

The fauna of Aruba contains no rain-forest species. For this reason I cannot agree with WINKLER's supposition (1926) that great forests existed on the island before the arrival of the Europeans. Compare in this connection the description of the flora in BOLDINGH (1914), HUMMELINCK (1933, 1938, 1939), and JOHNSTON (1909). The geological and palaeontological data as yet available cannot be readily interpreted for Aruba. For the present they do not clearly indicate whether or not the island had a land connection to the continent during the Pleistocene as a result of a eustatic fall in sea level (cf. VAUGHAN 1901, 1919, RUTTEN 1932, WESTERMANN 1932, EMILIANI 1955, ALEXANDER 1961, BUISONJÉ 1964).

The close relationship between the Barranquilla and Magdalena centres indicates a faunal exchange in previous times between the two centres (see discussion under Cauca and Magdalena centres).

9. The Santa Marta centre

The distribution of the faunal elements of the centre is dependent on the occurence of isolated lowland rain forests at the foot of the Sierra de Santa Marta (5776 m.). Such faunal elements include *Coendou bicolor sanctaemartae, Heteromys anomalus jesupi, Proechimys guyannensis nincae* (Rodentia) and other forms (fig. 5, no. 9). At the present time the centre is screened off from the Nechi subcentre of the Colombian Pacific centre by the dry areas south of the Sierra (see under Colombian Pacific centre).

Only one faunal element of the birds is specifically distinct (*Crax annulata*). On the other hand the number of monocentric subspecies is high. Thus 37 birds, 12 mammals, nine reptiles and seven amphibia are subspecifically distinct, although *Eupemphix pustulosus ruthveni* and *Basiliscus basiliscus barbouri* seem also to occur in the montane forest zone. The Sierra Nevada centre, which lies above the Santa Marta centre in the montane forests, has by comparison a much higher proportion of endemic species.

Among the birds, the faunal elements of the Santa Marta centre include no endemic Colombian species and belong in 81% of cases to widely distributed *Formenkreise*, which also occur in the Central American centres and in the Guyanan centre. Such forms include: *Tinamus major zuliensis, Urubitornis solitaria solitaria, Ortalis ruficauda ruficrissa, Laterallus melanophaius cerdaleus, Geotrygon linearis infusca, Lurocalis semitorquatus semitorquatus, Caprimulgus rufus rufus, Threnetes ruckeri darienensis, Phaethornis superciliosus susurrus, Chalybura buffoni aeneicauda, Anthocephala floriceps floriceps, Pharomachrus fulgidus festatus, Trogon violaceus caligatus, Aulacorhynchus prasinus lautus, Piculus rubiginosus alleni, Xiphocolaptes promeropirhynchus sanctaemartae, Dendrocolaptes picumnus seilerni, Anabacerthia striaticollis anxius, Automolus rubiginosus rufipectus, Sakesphorus canadensis phainoleucus, Myrmotherula schisticolor sanctaemartae, Formicarius analis virescens, Grallaria rufula spatistor, Attila spadiceus parvirostris, Terenotriccus erythrurus fulvigularis, Tolmomyias sulphurescens exortivus, Ornithion semiflavum dilutum, Microcerculus marginatus corrasus, Cyanerpes caeruleus caeruleus, Chlorophonia cyanea psittacina.*

The faunal elements of the Santa Marta centre have close affinities in the forest biomes north and south of the Sierra de Perija and of the central part of the Magdalena valley. However the endemic rain-forest forms that occur in these biomes are at most subspecifically distinct and make up less than 2% of the total fauna. These areas cannot therefore be seen as true centres equivalent in value to the Santa Marta centre or the Nechi subcentre of the Colombian Pacific centre.

10. The Sierra Nevada centre

The centre (fig. 5, no. 10) can be defined on the ranges of *Oryzomys villosus*, *O. albigularis maculiventer* (Rodentia, Cricetidae), *Eleutherodactylus carmelitae*, *E. delicatus*, *E. insignitus*, *E. megalops*, *E. sanctaemartae* (Anura, Leptodactylidae) and *Geobatrachus walkeri* which belongs to a monotypic Leptodactylid genus. The localities where the faunal elements have been found occur without exception above 1500 m. in the montane forests of the Sierra Nevada.

Among the 112 breeding birds found in the centre 72 (i.e. 65%) are subspecifically distinct (according to MEYER DE SCHAUENSEE, 1964). Fifteen birds constitute good species, i.e.: *Crypturellus idoneus, Pyrrhura viridicata, Campylopterus phainopeplus* (which according to MEYER DE SCHAUENSEE (1964, 1966) is restricted to between 1400 and 1800 m. from February to May, but from June to October goes up to 4500 m.), *Anthocephala floriceps* (*Anthocephala* being a monotypic Trochilid genus), *Coeligena phalerata, Synallaxis fuscorufa, Cranioleuca hellmayri Grallaria bangsi, Myiotheretes pernix, Myioborus flavivertex, Basileuterus basilicus, Conirostrum rufum, Anisognathus melanogenys, Catamenia oreophila, Atlapetes melanocephalus*. These 15 species have their closest affinities among the montane forest populations of the Colombian Eastern Cordillera. *Crypturellus idoneus* and *Basileuterus basilicus* can be derived from Central American montane forest populations. *Myioborus flavivertex* has its closest relatives in the Venezuelan montane forest centre.

The herpetofauna of the centre has not yet been sufficiently analysed. As a result of personal re-examination I am convinced of the specific status of *Eleutherodactylus delicatus, E. insignitus, E. megalops* and *E. sanctaemartae*. Examination of the holotype of *Geobatrachus walkeri* also confirmed the validity of this monotypic Leptodactylid genus (RUTHVEN 1915, Occ. Pap. Mus. Zool. Univ. Michigan, Ann Arbor, 20, p. 2, DUNN 1944, Caldasia 210, p. 519, GRIFFITHS 1959, *Proc. Zool. Soc. London* 1959, p. 480–481). In addition to this it is possible to affirm the validity of three Iguanids and two Teiids which must be taken as faunal elements of the centre i.e. *Anolis gaigei, A. solifer, A. solitarius, Anadia pulchella* and *Bachia talpa*.

11. The Magdalena centre

The position of the centre (fig. 5, no. 11) can be defined on the ranges of: *Sylvilagus floridanus purgatus* (Lagomorpha), *Sciurus granatensis variabilis* (Rodentia, Sciuridae), *Campylorhynchus griseus bicolor* (Aves, Troglodytidae), *Arremon conirostris inexpectatus* (Aves, Fringillidae), and *Speotyto cunicularia tolimae* (Aves, Strigidae). It is the bird fauna of this centre which is best known and which has been exhaustively described by MILLER (1947, 1952), MEYER DE SCHAUENSEE (1964) and HAFFER (1967). Of the 192 bird species which occur in the whole Magdalena graben, 54 require an unforested biotope. Of these 54 species, 23 are subspecifically distinct. There are no endemic species.

Relict rain forests existed in the central part of Magdalena graben even during the post-glacial arid phase and acted as barriers to the spread of species adapted to an open landscape (cf. Discussion). This can be deduced from certain endemics of these islands of forest (*Anolis stigmosus, Euphonia trinitatis*). These endemics are too small in number, however, to justify considering these areas as an independent centre of general significance for all terrestrial rain-forest vertebrates.

The 23 subspecifically distinct birds of the Magdalena centre immigrated from the north (Barranquilla centre), as shown by the following table:

Barranquilla & Caribbean Centres	Magdalena Centre
Lepidopyga goudoti	*L. g. goudoti*
Myrmeciza longipes	*M. l. boucardi*
Campylorhynchus griseus	*C. g. zimmeri*
Burhinus bistriatus	*B. b. vocifer*
Speotyto cunicularia	*S. c. tolimae*
Chordeiles acutipennis	*C. a. crissalis*
Nystalus radiatus	*N. r. radiatus*
Thamnophilus doliatus	*T. d. albicans*
Formicivora grisea	*F. g. hondae*
Manacus manacus	*M. m. flaveolus*
Idioptilon margaritaceiventer	*I. m. septentrionalis*
Basileuterus rivularis	*B. r. motacilla*
Tiaris bicolor	*T. b. huilae*
Coryphospingus pileatus	*C. p. rostratus*
Arremon conirostris	*A. c. inexpectatus*
Ortalis ruficauda	*O. guttata columbiana*
Ramphocelus dimidiatus	*R. d. molochinus*
Sporophila intermedia	*S. i. agustini*
Colinus cristatus	*C. c. leucotis*
Columbigallina passerina	*C. p. parvula*
Catharus aurantiirostris	*C. a. insignis*
Polioptila plumbea	*P. p. anteocularis*
Falco sparverius	*F. s. intermedius*

Fossils found up to now indicate that the centre was inhabited during the Miocene and Pliocene by a hygrophile fauna (ESTES 1961, ESTES & WASSERSUG 1963, SAVAGE 1951).

12. The Cauca centre

This centre lies between the West and Central Cordilleras of Northern Colombia and thus includes the Cauca valley and the Patia valley which adjoins it to the south (fig. 5, no. 12). The two valleys are separated from each other by the Popayan plateau which is 1750 m. high. Of the 30 species of birds which live in the centre, 15 can be considered as faunal elements of it. These are: *Ortalis guttata caucae, Colinus cristatus badius, Zenaida auriculata caucae, Columbigallina passerina nana, C. talpacoti caucae, Leptotila verreauxi decolor, Forpus conspicillatus caucae, Myiopagis viridicata, Catharus aurantiirostris phaeopleurus, Falco sparverius caucae, Rallus nigricans caucae, Phyllomyias griseiceps caucae, Polioptila plumbea daguae, Cypseloides lemosi, Picumnus granadensis.*

It can be shown that these 15 faunal elements immigrated from the north, as for the Magdalena centre. Most of the species which got into the Cauca valley have also reached the Patia valley. Among reptiles, amphibia and mammals the disjunct populations of the two valleys are not subspecifically distinct. Thus the opossum *Caluromys derbianus*, which likewise reached the Cauca centre from the north, occurs as the same subspecies both in the Cauca valley and in the Patia valley. The same is true for the subspecies *S. g. valdiviae* of the squirrel *Sciurus granatensis*. Among birds only *Forpus conspicillatus pallescens* and *Thraupis virens quaesita* show morphological differentiations in the Patia valley which justify subspecific separation from the Cauca populations.

The occurrence in the Patia valley of *Myiopagis viridicata implacens* and *Saltator albicollis flavidicollis* is worth mentioning. Both species are absent in the Cauca valley but nonetheless occur in the Ecuadorian subcentre of the Andean Pacific centre. Since they occur as the same subspecies both in the Patia valley and in the coastal area of western Ecuador it is likely that they only reached the Patia valley very recently.

The low degree of differentiation between the Patia and Cauca populations can be explained by two factors. Firstly the plateau of Popayan does not present a sharp barrier to species requiring an open landscape; and secondly the forest biome which existed in the southern part of the Cauca valley obviously had no isolating effect in post-glacial times. (In historical times these relict forests have been largely destroyed by man.) The fact that forests were at one time more extensive can be deduced from the relict occurrences of the rain-forest species *Pipra erythrocephala, Formicarius analis* and *Manacus vitellinus*; however it is possible that these three species have already been exterminated in the Cauca valley, since all the locality data available are more than 36 years old. It is likely that the Popayan plateau will not at present permit a regular exchange of the lowland fauna of the two valleys that it separates. Seeing that observations from this plateau are at present entirely lacking we can only suspect that such an exchange sometimes occurs from the fact that the populations in the two valleys are so little different from each other. HAFFER's supposition (1967, p. 326), that the Popayan plateau could only be crossed during warm interglacial periods, is unlikely in view of this low degree of differentiation.

13. The Colombian montane forest centre

This centre (fig. 5, no. 13) can be defined on the ranges of: *Sciurus pucherani* (Rodentia, Sciuridae), *Nothocercus julius* (Aves, Tinamidae), *Atlapetes melano-cephalus* (Aves, Fringillidae), *Tantilla longifrontalis* (Reptilia, Colubridae), *Tantilla fraseri* (Colubridae) and *Atractus occidentalis* (Colubridae) (fig. 14).

Fig. 14. Distribution of *Nothocercus julius*, a faunal element of the Colombian montane forest centre. The species closest related to it is *N. nigrocapillus*, a Yungas faunal element. The third species of *Nothocercus* (*N. bonapartei*) is polycentric.

The localities where its faunal elements have been found lie in the montane forest biomes of Colombia (except for the Sierra de Santa Marta), Ecuador and western Venezuela.

The centre can be readily divided into two subcentres—the West Andean and the East Andean. The populations of the Central Andes are more closely related to those of the West Andean subcentre than to those of the East Andean subcentre. For this reason I include the Central Andes with the West Andean subcentre.

CHAPMAN (1926) was the first to analyse the bird fauna of the Colombian montane forest centre, and most of what he said on the subject is still valid.

Considering first the Ecuadorian montane forest biome of the centre, it appears that 272 species and subspecies of birds occur which can be considered as typical montane forest forms. 224 of these live on the east side of the Andes and 221 on the west side. Of the 221 that occur on the west side, 46 do not occur on the east (cf. HAFFER 1967). Of the 224 that occur on the east side, 49 do not occur on the west. 175 forms live on both sides with 101 as the same species and 74 as the same subspecies.

Extending these results into the Colombian area, where we can supplement them with the researches of MEYER DE SCHAUENSEE (1948–52, 1964), it appears that West Andean populations agree with those of the Central Andes in more than 81% of cases (= 117 species). On the other hand, of the 176 bird species cited by CHAPMAN (1917), MAYR & PHELPS (1967) and MEYER DE SCHAUENSEE for the montane forest of the Colombian Eastern Cordillera, only 84 (= 48%) occur in the Central Andean montane forest. This indicates that the isolation between East Andean and Central Andean montane forest populations of birds is, or was, more effective than between the West and Central Andean populations.

These results seem to be confirmed among the amphibia, reptiles and mammals. Of the 165 snakes of Ecuador, 31 should be regarded as faunal elements of the Choco subcentre of the Colombian Pacific centre, 12 as of the Andean Pacific centre and 41 as of the Amazon centre. 34 species are distributed polycentrically in Amazonia, while 22 species cannot yet be assigned to a dispersal centre, because their distribution is not well enough known. Eight species are to be taken as elements of the West Andean and 11 as elements of the East Andean subcentre of the Colombian montane forest centre. Of the 11 East Andean species 9 show close relationships to the Venezuelan montane forest centre and 2 to the Central Andean montane forest biome. Of the 8 West Andean species, 5 have their closest relatives in the Central Andean montane forests and 3 in the Central American montane forest centre, as compiled from the list given by PETERS (1960).

The relief and vegetational zonation of the area make it likely that during the Würm glacial period gene exchange took place between the West Andean and Central Andean populations by way of the Nechi subcentre (cf. Discussion).

According to the researches of GORHAM (1966) and SOLANO (1969) 6 species of *Eleutherodactylus* (Anura, Leptodactylidae) should be assigned to the Colombian montane forest centre. These are: *Eleutherodactylus affinis*, *E. briceni*, *E.*

buergeri, E. carrioni, E. crucifer, E. devillei. The localities for these species of amphibia lie higher than 1400 m.

Certain species have a distribution that largely agrees with that of *Nothocercus julius*, which is a faunal element of the Colombian montane forest centre (cf. fig. 14). Such are the two *Bothrops* species, *B. alticolus* and *B. xantogrammus*, as well as five Iguanids i.e. *Anolis andinus, A. apollinaris, A. frenatus, A. jacare* and *Leiocephalus formosus* (two males received from Mrs. NAUNDORF of Quito were collected at altitudes of 1540 and 1800 m. respectively).

The position of seven species of snake is still uncertain. These are: *Leptotyphlops affinis, Atractus erythromelas, A. ventrimaculatus, Dendrophidion*

Fig. 15. Polycentric montane forest species which indicate the relationship between Colombian montane forest centre and the Yungas centre.

36

percarinatum, Dipsas latifrontalis, Leimadophis bimaculatus opisthotaenia and *Tropidodipsas perijanensis*. They may be elements of the Colombian montane forest centre or may represent polycentric faunal elements. These seven species penetrate far to the north along the Colombian Eastern Cordillera, perhaps reaching the Venezuelan montane forest centre (at Guanare, Cordillera de Mérida).

The affinities of the faunal elements of the Colombian montane forest centre with the Pantepui centre can be read from fig. 28 (see also discussion of Pantepui centre). The degree of relationships with the Yungas centre will be discussed under that centre.

14. The Colombian Pacific centre

The limits of the faunal elements of this centre (fig. 5, no. 14) are correlated with the limits of the lowland rain forests which stretch from Western Ecuador in the south to the southern outliers of the mountains of Darien in the north (fig. 16, 17). Eastwards its faunal elements penetrate into the rain-forest biome of the northern Cauca and Magdalena grabens, but are lacking in the Cauca and Magdalena centres.

The subspecific distinctness of innumerable rain-forest species in the northern Cauca area, on the one hand, and in western Colombia and Ecuador on the other, justifies regarding the two areas as subcentres (cf. CHAPMAN 1926, MEYER DE SCHAUENSEE 1964, HAFFER 1967). The subcentre between the Rio Sinu and the lower reaches of the Rio Cauca I refer to as the Nechi subcentre, while the subcentre west of the Andes is the Choco subcentre. These names follow the terminology of CHAPMAN (1926) and HAFFER (1967, 1969).

The way in which the birds of the Colombian Pacific centre have been studied should serve as a model for the whole of South America (CHAPMAN 1917, 1926, WETMORE 1941, 1946, 1953, MEYER DE SCHAUENSEE 1950, 1948–52, 1964, 1966, DUGAND 1952, BLAKE 1955, 1962, OLIVARES 1958, 1962, SIBLEY 1958, HAFFER 1959, 1967, BORRERO, OLIVARES & HERNANDEZ 1962, HAFFER & BORRERO 1965). I shall therefore consider the bird fauna first and then compare it with what is known about the amphibia and reptiles (BOULENGER 1882, PERACCA 1914, RUTHVEN 1914, CORRINGTON 1929, WERNER 1916, AMARAL 1935, PARKER 1938, NICEFORO 1942, DUNN 1944, PRADO 1945, DANIEL 1949, SCHMIDT 1955, PETERS 1960, PETERS & OREJAS-MIRANDA 1970, TAMSITT 1964). As with the other centres I shall also take the mammals into account (CABRERA, 1957).

There are 1556 Colombian species of birds including migrants (MEYER DE SCHAUENSEE 1964, 1966), and 332 of them occur in the Colombian Pacific centre. Of these at least 40% (131 species) can be reckoned as faunal elements of the centre, including 5 monotypic genera (cf. fig. 16). Besides these there are a great number of polycentric species represented by monocentric subspecies (see below). The varying levels of differentiation of the faunal elements suggest that this differentiation started at very different times for different forms. To confirm this supposition it is necessary to look into the origins of the Choco and Nechi faunal elements.

In this connection, the origin of the monotypic genera *Androdon* and *Sapayoa* is still quite unclear. The monotypic Piprid *Allocotopterus* can be seen as related to *Machaeropterus regulus*. The latter species is found on the one hand in the coastal rain forests of Venezuela and the rain forests in the northerly Magdalena region as far as Caldas; on the other hand it occurs in the Amazon and Serra do Mar centres. The monotypic Thraupid genus *Erythrothlypis* is connected with the genus *Thlypopsis*, which is widespread in South America, and also with the Central American species *Chrysothlypis chrysomelas*. The Formicariid genus *Sipia*, known as two species (cf. fig. 16), is related to *Cercomacra* which is

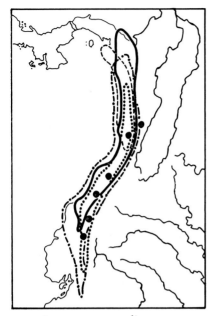 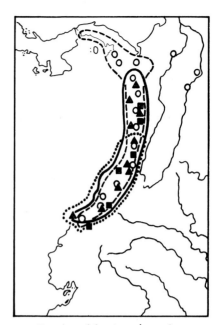

— Dipsas temporalis
····· Penelope ortoni
--- Eleutherodactylus bufoniformis
—·— Pteroglossus sanguineus
  ● Anaea pandrosa

— Erythrothlypis salmoni
--- Sapayoa aenigma
▲ Sipia berlepschi
····· Sipia rosenbergi
○ Androdon aequatorialis
■ Allocotopterus deliciosus

Fig. 16. Eleven faunal elements of the Colombian Pacific centre.

widespread in South America (cf. MEYER DE SCHAUENSEE 1966). Since fossils are completely lacking it is not possible to state what immigration routes the monotypic genera followed. Equally it would be purely hypothetical to suppose that these genera arose in the Colombian Pacific centre (CHAPMAN 1926).

On the other hand, an analysis of origins based on the present-day distributions of the species of polytypic genera gives firmer results. Three groups can be distinguished according to origin and will now be discussed.

The representatives of Group I reached the Colombian Pacific centre from the Amazon centre (no. 25) by crossing over the Andes of Peru and Ecuador. Possible immigration routes were discussed by CHAPMAN (1923), MILLER (1952), and KOEPCKE (1961). The most likely route for a number of forms was by way of the 2150 m. high Porculla Pass (Marañon) in northern Peru and the Loja route in southern Ecuador (CHAPMAN 1923). Such forms include *Sittasomus griseicapillus aequatorialis, Pachyrhamphus spodiurus, Trogon melanurus mesurus, Formicarius nigricapillus, Manacus manacus* and *Cacicus cela flavocrissus*. The

39

ranges of many other forms should also be compared in this connection; such are *Osculatia saphirina* (fig. 17), *Crax rubra*, *Attila cinnamomeus torridus*, *Micromonacha lanceolata*, *Pipra mentalis*, *Gymnopithys leucaspis*, *Thamnistes anabatinus*, *Microrhopias quixensis* (MEYER DE SCHAUENSEE 1964).

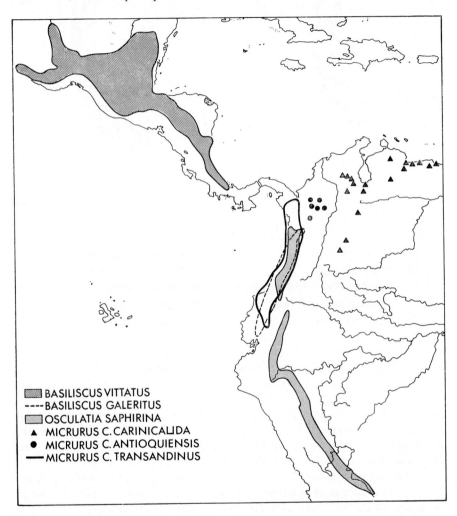

Fig. 17. Affinities of the Choco faunal elements *Basiliscus galeritus*, *Osculatia saphirina* and *Micrurus carinicauda transandinus*. The species closest relayed to *B. galeritus* is *B. vittatus*—a faunal element of the Central American rain-forest centre. *Micrurus carinicauda antioquiensis* is a Nechi faunal element.

Some of the populations in Group I are replaced by vicarious species in north-west Venezuela and north-east Colombia so that their passage over these areas would have been blocked. For others, the part of their range east of the

40

Andes reaches a northern limit south of the Rio Marañon (Ucayali subcentre of the Amazonian centre).

The assumption that Group 1 forms reached the Colombian Pacific centre by crossing the Andes explains the cis-trans-Andean disjunction of their ranges. Taxa with distributions of this type are in all cases absent from the Nechi subcentre and the Santa Marta centre. The possibility that such forms differentiated before the Andes arose can be excluded (cf. Discussion). Pre-glacial wanderings will not be discussed here because individual authors have reached such contradictory conclusions (cf. STUTZER 1925, CHAPMAN 1926, SCHÖNFELD 1947). The small degree of differentiation of the cis- and trans-Andean populations makes it unlikely that Group 1 forms crossed the Andes in pre-glacial times, even supposing that such a crossing was at that time necessary. If the systematic results are correlated with geological and palaeontological results it seems likely that the Andean immigration route of Group 1 forms was still used in post-glacial times [Older Peron (FAIRBRIDGE 1962)]. Relevant here are the works of: SCHMIDT (1952), WILHELMY (1954, 1957), BELDING (1955), WEYL (1956), MORTENSEN (1957), WEST (1957), VAN DER HAMMEN & GONZALES (1960), BÜRGL (1961) and VAN DER HAMMEN (1961). GRISCOM's view (1932, 1935) that the Choco area was settled entirely from the north must be mistaken, at least as far as Group 1 forms are concerned.

The representatives of Group 2 reached the Colombian Pacific centre from the Amazon centre, or at least from some centre in Amazonia. They came by way of the rain forest of the Venezuelan littoral and the Catatumbo and Santa Marta centres, and a migration route in the opposite direction can likewise be observed. The species and subspecies ascribed to this group show a degree of differentiation comparable to that of Group 1.

An example of Group 2 is *Pipra erythrocephala* which inhabits the forest areas of the Amazon centre (Napo subcentre), Venezuelan coastal forest centre, Santa Marta centre and the Nechi subcentre; an isolated population also occurs in the Gulf of Uraba. Another good example of Group 2 is the Elapid *Micrurus carinicauda* (cf. fig. 17). The nominate form is found in the Venezuelan coastal forest centre—three isolated occurrences are known from the eastern edge of the Colombian Eastern Cordillera—and the Catatumbo centre. In the Nechi subcentre the subspecies *M. c. antioquiensis* occurs and in the Choco subcentre the subspecies *M. c. transandinus*.

The representatives of Group 3 reached the Choco and Nechi subcentres from Central America (22 bird species). This group has already been noticed in discussing the Costa Rican centre. In individual cases, because of the lack of clinal variability, it is not always easy to say in which direction the migration moved. Indeed a number of forms that have their closest related species or subspecies in the East Andes suggest that dispersal outwards from the Choco subcentre has occurred. In this way the following species which occur in Central American can be defined as Choco faunal elements: *Leucopternis semiplumbea*, *Nystalus radiatus*, *Mitrospingus cassinii*, *Malacoptila panamensis*, *Myrmotherula fulviventris*, *Hylophylax naevioides*, *Lipaugus unirufus*, *Rhytipterna holerythra*,

Laniocera rufescens, Pachyramphus cinnamomeus, Celeus loricatus. On the other hand a number of species can be ascribed to Central American centres and spread into the Colombian Pacific centre from Central America. Such are: *Rhynchortyx cinctus, Hylomanes momotula, Oncostoma cinereigulare, Gymnocichla nudiceps, Myrmornis torquata* and *Selenidera spectabilis.* The present-day ecological situation allows Colombian Pacific faunal elements to wander northwards into the Costa Rican centre along the outlying hills of Darien, and also allows movement in the opposite direction.

The three indicated routes of immigration have obviously been used separately by individual taxa at varying times. This repeated immigration shows itself in present-day distributions when, in the same distribution centre, a relatively young immigrant occurs sympatrically with an older immigrant of the same genus. A comparison can be drawn with the settlement of islands (MAYR 1942, 1963, MÜLLER 1969, SAVAGE 1968, STRESEMANN 1927–1934, 1939). A series of such examples can be given.

Thus the Central American Iguanid genus *Basiliscus*, according to SAVAGE (1966), occurs in the Choco subcentre as a readily diagnosed species *Basiliscus galeritus* (fig. 17). In the northern part of the Choco region *B. galeritus* is found sympatrically with *Basiliscus basiliscus basiliscus*, which obviously reached this area from the Costa Rican centre. In the Santa Marta and Catatumbo centres there is a further subspecies of *B. basiliscus* (i.e. *B. b. barbouri*) while the Central American rain-forest centre is inhabited by an endemic species *Basiliscus vittatus* (fig. 17). Presumably *B. galeritus* and *B. basiliscus* are of monophyletic origin (MATURANA 1962) and *B. galeritus* reached the Choco subcentre long before *B. basiliscus.* The populations of *B. basiliscus* which later invaded the region during a second phase of expansion encountered a reproductively isolated population there.

Among the birds HAFFER (1967) has established the probability of double invasions (e.g. *Columba subvinacea, Tangara schrankii*) and also of a triple invasion (*Trogon melanurus*).

Among the amphibian faunal elements of the centre the high proportion of *Eleutherodactylus* species is particularly striking. Of the 18 Leptodactylid species whose limits can be correlated with those of the centre, as many as 14 are species of *Eleutherodactylus.* These are: *Eleutherodactylus achatrinus, E. anomalus, E. biporcatus, E. bisignatus, E. bufoniformis, E. cornutus, E. diastema, E. fitzingeri, E. latidiscus, E. longirostris, E. palmatus, E. roseus, E. taeniatus, E. w-nigrum.* Monotypic genera are lacking among amphibia, both in the anurans and the urodeles. Affinities are closest with the Costa Rican, Santa Marta and Amazon centres (cf. also fig. 18, 19).

A close relationship with the Costa Rican centre is indicated above all by the species of *Eleutherodactylus* and *Dendrobates* (cf. COCHRAN 1966, SAVAGE 1968). For certain species of Leptodactylid which must be considered as faunal elements of the Colombian Pacific centre, the phylogeny is not yet sufficiently clear. These are: *Eleutherodactylus conspicilatus, E. raniformis, E. brederi, Ceratophrys stolzmanni, Eupemphix pustulatus, Leptodactylus ventrimaculatus*

42

Fig. 18. Distribution of two very closely related species of *Bolitoglossa. B. peruviana* is an Amazon faunal element and *B. chica* is a Choco faunal element.

and *Leptodactylus wagneri*. Colombian pacific elements are: *Eleutherodactylus bufoniformis, E. conspicillatus, E. anomalus, E. raniformis, E. longirostris, E. brederi* and *Eupemphix pustulosus* (cf. COCHRAN, D. M. and GOIN, G. J. 1970, Smiths. Inst. Bull. 288).

The Iguanid genus *Anolis* also includes at least 23 species which must be reckoned as faunal elements of the Colombian Pacific centre. These are: *Anolis aequatorialis, A. binotatus, A. biporcatus parvauritus, A. bitectus, A. chloris, A. chocorum, A. festae, A. eulaemus, A. fasciatus, A. fraseri, A. gemmosus, A. lemniscatus, A. gorgonae, A. peraccae, A. gracilipes, A. graniculeps, A. macrolepis, A. latifrons, A. mirus, A. p. pentaprion, A. palmeri, A. ventrimaculatus, A. princeps.*

43

Fig. 19. Distribution of the species *Bolitoglossa altamazonica* and *B. sima* (after BRAME and WAKE 1963). Within the group of *B. amazonica* these two species are closest related. *B. sima* is a Choco faunal element. *B. altamazonica* is a polycentric faunal element which, outside the Amazon centre, is known from an isolated population from the middle Magdaelena graben and from the Para centre. The species is lacking in the region of REINKE's savanna corridor (1962, cf. fig. 34 herein).

As against this the Teiids, which are so widely distributed in South and Central America, are represented only by 5 monocentric faunal elements in the Colombian Pacific centre and the Crotalids only by two. The Teiids are: *Anadia rhombifera, A. vittata, A. angusticeps, Echinosaura horrida palmeri* and *Prionodactylus vertebralis*. The Crotalids are: *Bothrops punctatus* and *B. microphthalmus colombianus*, the other subspecies of *B. microphthalmus* being a faunal element of the Amazon centre.

15. The North Andean centre

This centre (fig. 5, no. 15) can be defined on the ranges of a number of forms, including: *Pudu mephistophiles, Sylvilagus brasiliensis andinus, Thomasomys paramorum, Anatomys leander, Nasuella olivacea, Marmosa dryas, Hemispingus*

PÁRAMOS
▲ CISTOTHORUS MERIDAE
● HEMISPINGUS VERTICALIS
△ PHENACOSAURUS RICHTERI

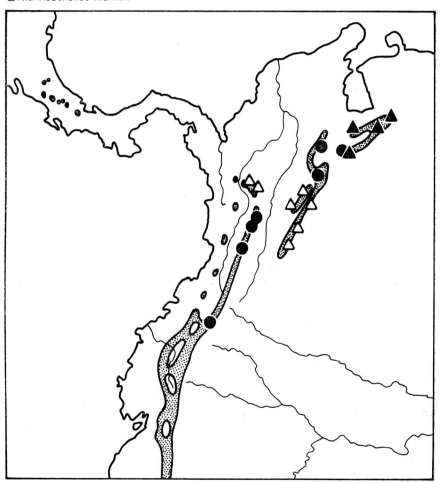

Fig. 20. The ranges of three faunal elements of the North Andean centre in relation to the distribution of the Colombian paramo.

verticalis, Cistothorus meridae, Schizoeca fuliginosa, Phenacosaurus richteri, Nothoprocta curvirostris, Chalcostigma olivaceum, Cinclodes excelsior, Schizoeaca griseomurina and *Asthenes virgata* (figs. 20, 22, 23). The faunal elements of the centre show a definite barrier-zone in southern Ecuador (cf. the distribution of the insectivore *Cryptotis* in fig. 21). Some mammals and birds nonetheless extend farther south into the paramo of northern Peru, west of the Marañon river, as far as Lake Junin, or else are represent there by subspecifically distinct populations or closely related species. I therefore regard the paramo area of northwestern Peru as a distinct subcentre (**Peruvian Andes subcentre**) as opposed to the **Bogotá subcentre** which takes in the High Andes of Ecuador, Colombia and Venezuela.

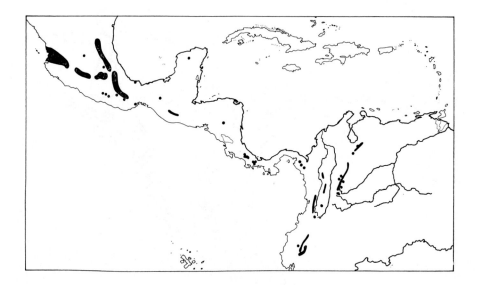

Fig. 21. Distribution of the insectivore genus *Cryptotis*. In Mexico and elsewhere in Central America species of this genus occur in the montane forests and paramos.

The Peruvian Andes subcentre has the character of a transition-zone between the Bogotá subcentre and the Puna centre. The more mobile taxa, in particular, connect the North Andean and Puna centres (cf. the ranges of *Tremarctos ornatus, Anthus bogotensis* and *Troglodytes solstitialis*).

The preferred biotope of the North Andean faunal elements is the paramo of the North and Central Andes above 3000 m. (cf. WEBER 1958, TROLL 1931). Locality data available for the birds (cf. CHAPMAN 1926) nonetheless show that the faunal elements, especially outside the breeding season, can be met with in the almost vegetation-less boulder areas, in the areas of bed rock and in the thorn cushion regions of the mountains. In isolated cases they even occur in the montane forest zone.

The strong orographic subdivision of the North Andean centre increases the degree of isolation of the populations. There are endemic species and subspecies in the paramos of the Sierra Nevada de Santa Marta (*Ramphomicron dorsale*) and the Sierra de Mérida (*Cistothorus meridae*); these indicate that the centre could be subdivided still further not merely into subcentres but at a tertiary or quaternary level. The subspecific differentiation of the range of *Nasuella olivacea* should be considered in this connection (fig. 22).

Some species among North Andean faunal elements have their closest relationships with populations of the Talamanca paramo centre, the Puna centre and the unforested mountain tops of the Roraima region (cf. fig. 24).

Fig. 22. Nasuella olivacea and *Marmosa dryas* as examples of North Andean faunal elements. Also shown are the ranges of *Tremarctos ornatus* and *Anthus bogotensis* to illustrate the relationship between the North Andean and the Puna centre.

47

Fig. 23. The ranges of: *Nothoprocta parvirostris*, a faunal element of the Peruvian Andes sub-centre of the North Andean centre; *Ramphomicron dorsale*, an endemic species of humming bird from the paramo of the Sierra Nevada; *Catamenia homochroa*—note the isolated occurrence in the Pantepui centre; *Asthenes flammulata*, a North Andean faunal element; and *Asthenes maculicauda*, a Puna faunal element.

Starting from the North Andean centre two species have thus succeeded in invading the Roraima region, though obviously at very different times. One of these is *Troglodytes rufulus* which can be derived from *T. solstitialis* (cf. fig. 24); it is represented by five subspecifically distinct populations in the unforested summit zone of the Pantepui centre i.e. *T. r. rufulus*, *T. r. wetmorei*, *T. r. duidae*, *T. r. yavii* and *T. r. fuligularis*. (cf. MAYR & PHELPS, 1967). The other species is

48

Fig. 24. The ranges of two species of *Troglodytes*. *T. solstitialis* is a polycentric species which indicates a close affinity between the Talamanca paramo centre, the North Andean centre and the Puna centre. The closest related species is *T. rufulus* which occurs in the unforested region of the Pantepui centre.

Catamenia homochroa, represented by the endemic Pantepui subspecies *C. h. duncani* (cf. fig. 23).

Before dealing with the regional origin of the paramo faunal elements— whether from South, Central or North America—I shall first list the faunal elements among the birds of the Bogotá and Peruvian Andes subcentres. Bogotá faunal elements are: *Phalcoboenus carunculatus* (Falconidae), *Eriocnemis vestitus* (Trochilidae), *Eriocnemus derbyi*(Trochilidae), *Ramphomicron dorsale* (Trochilidae), *Metallura williami* (Trochilidae), *Chalcostigma heteropogon* (Trochilidae),

49

Oxypogon guerinii (Trochilidae), *Cinclodes excelsior* (Furnariidae), *Schizoeaca fuliginosa* (Furnariidae), *S. griseomurina* (Furnariidae), *Cistothorus apolinari* (Troglodytidae), *C. meridae* (Troglodytidae), *Hemispingus verticalis* (Thraupidae), *Catamenia haemorhoa* (Fringillidae), *Spinus spinescens* (Fringillidae). Peruvian Andes faunal elements are: *Nothoprocta curvirostris* (Tinamidae), *Phalcoboenus megalopterus* (Falconidae), *Metriopelia melanoptera* (Columbidae), *Oreotrochilus estella* (Trochilidae), *Chalcostigma olivaceum* (Trochilidae), *C. stanleyi* (Trochilidae), *Leptasthenura andicola* (Furnariidae), *Asthenes virgata* (Furnariidae), *Phacellodromus striaticeps* (Furnariidae). Species which occur in both subcentres are: *Metallura tyrianthina* (Trochilidae), *Asthenes flammulata* (Furnariidae), *Agriornis montana* (Tyrannidae), *Muscisaxicola alpina* (Tyrannidae), *Myiotheretes erythropygius* (Tyrannidae), *Troglodytes solstitialis* (Troglodytidae), *Anthus bogotensis* (Motacillidae), *Diglossa carbonaria* (Coerebidae), *Catamenia analis* (Fringillidae), *Catamenia inornata* (Fringillidae), and *Phrygilus unicolor* (Fringillidae).

According to MAYR (1964) only two of the families to which the faunal elements belong can be regarded with high probability as South American in origin. These are the Furnariidae and the Tinamidae. Accordingly only three species in the Peruvian Andes subcentre and four in the Bogotá subcentre belong to families that originated in South America. All the others are of North or Central American origin.

If the mammals are considered from the same viewpoint it appears that 28 species of the 37 North Andean faunal elements belong to families of North or Central American origin i.e. more than 75%. This is based on data from SIMPSON (1950, 1969), CABRERA (1957), HERSHKOVITZ (1966) and MANN (1968). Although the mammals of this area, like the amphibia, have not been satisfactorily worked from the systematic aspect, no great alteration in this percentage is to be expected.

The reptiles play a subordinate role in the North Andean centre on account of the prevailing climate. The Iguanid *Phenacosaurus* is represented by three species in the Colombian paramo i.e. *P. heterodermus, P. richteri*, and *P. nicefori. Phenacosaurus heterodermus*, which is the commonest species in the paramos of Bogotá, has a prehensile tail like the other species and prefers to live in bushes of *Espeletia, Rubus* and *Chuswquea*. Judging by its anatomy and ethology (KÄSTLE 1965) *Phenacosaurus* can be derived from the species of *Anolis* of the tropical lowlands and may share common ancestors with *Norops* of northern South America and Panama.

Some species of the Microteiid genus *Euspondylus* also seem to be North Andean faunal elements. This is certainly true for *Euspondylus brevifrontalis* (cf. FOUQUETTE, 1968).

The herpetofauna of Loja (Ecuador), was described by PARKER (1938) who also gave some information on geographical ranges and vertical distribution. Unfortunately, however, no further information has become available since that time (cf. PETERS 1960). The Loja fauna would be of the greatest interest for working out the migration routes between the Colombian Pacific centre and the Amazon centre. I have therefore tried to obtain material through the friendly offices of Mrs. NAUNDORF of Quito, but unfortunately in vain. For the present, therefore, the amphibia and reptiles of Loja cannot be considered.

Within the amphibia, especially the family Leptodactylidae, seven species can be reckoned as faunal elements of the North Andean centre. These are: *Eleutherodactylus bogotensis, E. elegans, E. flavomaculatus, E. lehmanni, Niceforonia nana, Telmatobius niger* and *T. vellardi.* A particularly interesting form is *Niceforonia nana* whose terra typica is the Paramo de La Russia. This species was only described in 1963, and belongs to a monotypic genus (GOIN & COCHRAN 1963, *Proc. California Acad. Sci.,* San Francisco 31, p. 500). Its phylogeny is still largely unknown.

The vegetation of the North Andean centre has been exhaustively described by GOEBEL (1891), WEBERBAUER (1911, 1945), HAUMAN (1918, 1931), ESPINOZA (1932), DIELS (1934) and WEBER (1958, 1969). Concerning adaptational mechanisms in the biota of the High Andes see ALTMANN & DITTMER (1961), DARLINGTON (1943) and MANN (1968).

16. The Catatumbo centre

The centre (fig. 5, no. 16) lies between the south-east slopes of the Sierra de Perija—i.e. the catchment of the Rio Catatumbo—and the north-west side of the Cordillera de Mérida.

The faunal elements ascribed to it penetrate along the Catatumbo into the Maracaibo basin. Some also pass along the lower limit of the montane forest in the east towards Motatan and in the north-west as far as Rosaria. On the other hand in the dry coastal region (cf. Caribbean centre) they are completely lacking (cf. fig. 25).

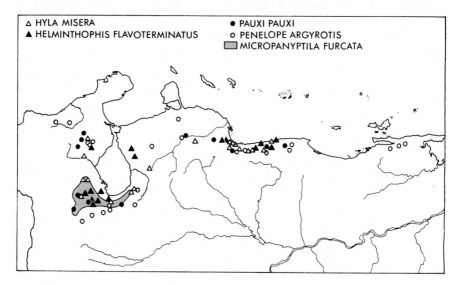

Fig. 25. Micropanyptila furcata—a monotypic Apodid genus—is a faunal element of the Catatumbo centre. *Hyla misera, Helminthophis flavoterminatus, Pauxi pauxi* and *Penelope argyrotis* are polycentric species which demonstrate the relationship between the Venezuelan coastal forest centre and the Catatumbo centre.

Some species with good powers of locomotion, as for example among bats and birds, occur as disjunct populations in the Venezuelan coastal forest centre, in the Catatumbo centre and also, more rarely, in the Santa Marta centre (e.g. *Pauxi pauxi* cf. fig. 25). In the absence of subspecific differentiation it is not possible to ascribe such species to a dispersal centre. This is true for some birds (e.g. *Spinus cucullatus*) amphibia (e.g. *Hyla misera*) and reptiles (e.g. *Helminthopis flavoterminatus* cf. fig. 25). These polycentric species do not argue against the independence of the Catatumbo centre. On the contrary, the well studied birds show its existence quite clearly; thus 21% of the rain-forest birds that occur in the centre are subspecifically distinct (MEYER DE SCHAUENSEE 1964,

HAFFER 1969). Moreover the occurrence of a monotypic genus of the family Apodidae is worthy of note (*Micropanyptila furcata* cf. fig. 25).

The low level of differentiation among reptiles and amphibia indicates that the present-day ecological isolation between the Catatumbo and Santa Marta centres, caused by the Sierra de Perija and the dry area of Guaira, is of recent origin. The same holds for the isolation between the Catatumbo and Venezuelan coastal forest centres. This statement is based for the reptiles on the works of: ALEMAN (1952, 1953), BOETTGER (1893), BOULENGER (1903, 1905), BRICENO (1934), BRONGERSMA (1940), FOWLER (1913), HELLMICH (1940, 1953), MILA DE LA ROCA (1932), PETERS (1877), ROHL (1949), ROUX (1927), ROZE (1952, 1959, 1966) and VELLARD (1941). For the amphibia it is based on: BRONGERSMA (1948), LUTZ (1927), PARKER (1933, 1936), RIVERO (1961) and WALKER & TEST (1955). Naturally the greater richness in endemic species of the Venezuelan coastal forest centre indicates that this is a more important dispersal centre than the Catatumbo centre (cf. also Santa Marta centre).

The differentiation of the bird fauna of the Catatumbo centre indicates that it has not been isolated long. Thus there are 37 subspecifically distinct populations among 176 species that occur, and there is one monotypic genus (cf. Discussion). The occurrence of a monotypic genus does not contradict this view, since no one can prove that *Micropanyptila furcata* actually arose in the Catatumbo centre.

17. The Venezuelan coastal forest centre

This centre (fig. 5, no. 17) lies in the north-west part of the coastal mountains between Caracas in the east and Tucacas in the north-west. It is only significant for the lowland fauna. Above 1400 to 1500 m. forms are found which must be regarded as montane forest species. These indicate the existence of an independent montane forest centre (fig. 5, no. 18).

Unlike the Santa Marta or the Catatumbo centre, the Venezuelan coastal forest centre has a large number of endemic species. Of the 612 Venezuelan bird species that occur here, 22 (3.6%) can be reckoned as faunal elements of the centre. This is based on data in BEEBE & CRANE (1948), FERNANDEZ YEPES (1946), MEYER DE SCHAUENSEE (1966), OSGOOD & CONOVER (1922), PHELPS (1938, 1943), PHELPS & PHELPS (1958–1963), and SCHÄFER (1969). The amphibia have about 12% of endemic species (12 species), based on data from RIVERO (1961, 1964), MERTENS (1950, 1957), SCHMIDT (1932), SHREVE (1947), SOLANO (1969), STEJNEGER (1901), TEST (1956) and WALKER & TEST (1955). This is roughly the same percentage as with the snakes, according to data from AMARAL (1927, 1929), DUNN (1928), MARCUZZI (1950), PIFANO (1935, 1938, 1954), PIFANO & ROMER (1949), PIFANO, ROMER & SANDNER (1950), ROZE (1953, 1954, 1955, 1958, 1959, 1966) and ROZE & TREBBAUM (1958). Closely similar percentages of endemics can be established for the invertebrates [cf. BAKER (1923) and others].

It is noteworthy in this connection that, among Lepidoptera, even such a widespread and strongly-flying Sphingid as *Protoparce florestan* is represented

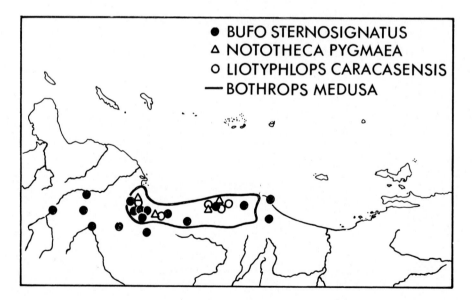

Fig. 26. Four characteristic faunal elements of the Venezuelan coastal forest centre.

by a particular subspecies in the Venezuelan coastal forest centre i.e. *P. f. vogli* (cf. DANIEL 1950).

There are also genetic connections with the Guyanan centre, in addition to those with the Catatumbo centre. However, among amphibians, of the 31 faunal elements of the Guyanan centre which extend into Venezuela only two occur in the coastal forest centre (cf. RIVERO 1961). The others are restricted in the Venezuelan area to the states of Amacupo and North-Bolivar. The relationship between the Venezuelan coastal forest centre and the Guyanan centre is stronger among the birds, corresponding to their greater powers of movement. Compare in this respect the discussion of Trinidad and Tobago under the Guyanan centre.

Some faunal elements of the Venezuelan coastal forest centre penetrate eastwards into the area of the Venezuelan montane forest centre (fig. 26). This penetration, however, roughly coincides with valleys running westward along the length of the mountain range. In this way two centres interdigitate at different altitudes in the Venezuelan Coast Range. The faunal elements of the one are ecologically limited to the lowland rain forest and of the other to the montane forest.

18. The Venezuelan montane forest centre

The position of the centre (fig. 5, no. 18) is determined by the distribution of the montane forests of the Venezuelan coastal Cordillera. The patterns of distribution of coastal forest elements and montane forest elements at first sight appear to coincide when their ranges are plotted on a map of Venezuela. However detailed analysis of the precise localities of Venezuelan snakes, for example, shows very clearly that the faunal elements of the Venezuelan montane forest centre only occur above 1400 m. while those of the Venezuelan coastal forest centre are only found below 1500 m. (data in ROZE, 1966).

In accordance with this boundary at 1400–1500 m. the following snakes should be considered as faunal elements of the Venezuelan montane forest centre: *Leptotyphlops macrolepis, Atractus badius, A. fuliginosus, A. univittatus, A. vittatus, Chironius monticola, Leimadophis zweifeli, Urotheca lateristriga multilineata, U. williamsi.* The vertebrate fauna of the Venezuelan coastal Cordillera has been well studied (BEEBE & CRANE 1948, PHELPS & PHELPS 1958–1963, RIVERO 1961, ROZE 1966). Not much attention, however, has been given to precise vertical distribution.

The close relationships between the Venezuelan montane forest centre and the Pantepui centre will be discussed under that centre (cf. also fig. 28).

Relationships with the Colombian montane forest centre are even stronger. *Leptotyphlops macrolepis, Atractus badius* and *A. fuliginosus*, which I take as faunal elements of the Venezuelan montane forest centre, are closest related to populations in the Colombian montane forest centre (*Leptotyphlops affinis, Atractus erythromelas* and *A. ventrimaculatus* respectively).

The passeriform birds are well suited to zoogeographic analysis and include 93 forms—species and subspecies—which must be seen as faunal elements of the centre. This constitutes over 50% of the passeriforms which occur in the Venezuelan coastal and montane forests. The remaining species are widespread in the Colombian East Andes. At the moment, however, no final appraisal can be made because the East Andean montane forest biome has not been properly studied. It is worth mentioning that four species of *Eleutherodactylus* (Amphibia, Leptodactylidae) must be considered as faunal elements of the centre, i.e. *Eleutherodactylus anotis, E. bicumulus, E. stenodiscus,* and *E. terraebolivaris.*

If the Venezuelan coastal and montane forest centres are compared with the Santa Marta and Sierra Nevada centres it emerges that the montane forest biome has a higher proportion of endemics. This is shown, for example, by the percentage figures for the avifauna.

19. The Caribbean centre

The limits of this centre (fig. 5, no. 19) are indicated by the limits of range of *Bothrops lansbergii venezuelensis, Pyrrhuloxia phoenicea* and *Columba corensis* (fig. 13).

Seven Venezuelan snake species must be considered as faunal elements of the centre i.e.: *Typhlops lehneri, Coluber mentovarius suborbitalis, Dryadophis amarali, D. bifossatus striatus, Erythrolamprus pseudocorallus, Oxyrhopus venezuelanus, Crotalus vegrandis. Crotalus durissus cumanensis* can also be taken as an expansive Caribbean faunal element. Starting from the nucleus of the centre in REINIG's sense (1950), as defined by the ranges of *Columba corensis* etc., this form has spread out into the humid savanna of the Llanos.

A distributional analysis of the 1282 Venezuelan bird species shows that the Llanos populations are so closely related to those of the arid coastal area of Venezuela that no Llanos centre can be delimited. This is based on data in PHELPS & PHELPS (1958–1963) and MEYER DE SCHAUENSEE (1966). Nevertheless the Rio Apure seems to have had an isolating effect on some populations. Thus the rabbit-owl *Speotyto cunicularia brachyptera* occurs in all unforested areas, but south of the Apure is represented by the subspecies *S. c. apurensis. Colinus cristatus continentis* inhabits the plains north of the Apure but south of it is replaced by *C. c. barnesi.*

Two subcentres can be recognised in the Caribbean centre. One of them coincides with the dry areas of the Guajira Peninsula and Falcon on the Gulf of Maracaibo and can be called the Maracaibo subcentre (cf. the range of *Pipa parva* in fig. 13). The other includes the dry Venezuelan coastal region east of the Paraguana Peninsula and can be called the Venezuelan subcentre.

The two subcentres can be defined by the ranges of their elements *Pipa parva* (fig. 13) and *Amazona barbadensis.* The Cordillera of Mérida and the Venezuelan coastal forest centre do not make an effective barrier in terms of present-day ecology. Moreover man has made artificial savannas which have led to the expansion of animals adapted to natural savannas (cf. OTREMBA 1954, WILHELMY 1954, VARESCHI 1955, MÜLLER 1970, 1971, 1972). Thus the savannas north of Caracas, which Vareschi sees as artificial, are populated by forms which arose in the Llanos and in the Caribbean coastal region and which are not distinct from the parent populations.

The island of Margarita can likewise be referred to the Caribbean centre. The herpetofauna of this island suggests a very short period of isolation. The degree of differentiation of populations on the island lies within the range of variation of the mainland populations. In the present state of knowledge the herpetofauna of the island consists of the following species: Amphibia; *Bufo marinus* (introduced), *Pleurodema brachyops.* Reptilia;

Gonatodes vittatus	*Tropidodactylus onca*
Phyllodactylus muelleri	*Tropidurus torquatus hispidus*
Iguana iguana iguana	*Ameiva bifrontata bifrontata*

Cnemidophorus lemniscatus lemniscatus
Tretioscincus bifasciatus
Leptotyphlops albifrons margaritae
Constrictor constrictor constrictor
Corallus hortulanus cookii
Epicrates cenchria maurus
Coluber mentovarius suborbitalis
Dryadophis amarali
D. pleei
Drymarchon corais margaritae
Leimadophis melanotus

Leptodeira annulata ashmeadii
Leptophis ahaetulla coeruleodorsus
Oxybelis aeneus aeneus
Phimophis guianensis
Pseudoba neuwiedii neuwiedii
Spilotes pullatus pullatus
Micrurus isozonus
Bothrops lansbergii venezuelensis
Crotalus durissus cumanensis
Crocodylus intermedius

The populations of *Leptotyphlops albifrons* and *Drymarchon corais* on Margarita show slightly divergent features from the mainland populations and ROZE (1966) gave these differences a subspecific value. Applying the 75% rule, however, the island populations cannot be clearly separated from those on the mainland.

The make-up of the herpetofauna shows quite clearly that Margarita has never been free of forests since its isolation from the mainland. The occurrence of the rain-forest species *Sibon nebulata*, *Oxybelis aeneus* and *Leptophis ahaetulla* can only be explained on the supposition of perennial rain forests on the island. This result is supported by the floristic studies of BEARD (1949). However the high proportion of species which must be taken as indicators of unforested areas shows that the island must be ascribed to the Caribbean centre. Thus of the 24 indigenous amphibia and reptiles, 16 must be regarded as species adapted to an open landscape. The same is true for a number of other islands off the Venezuelan coast within the 200 m. isobath. The islands of Aruba, Tortuga, Cubagua, Los Testigos, Margarita, Los Frailes, Isla de Caribes and Isla Coche would have climax vegetations generally dominated by forest, but unlike Trinidad and Tobago mainly by semi-deciduous forest. The bird and mammal faunas support this view, but are more strongly differentiated than the herpetofauna (cf. ALLEN 1902, BANGS 1898, LOWE 1907, MILLER 1898). Thus *Aratinga pertinax*, a characteristic bird of the unforested landscapes of South America (cf. MÜLLER 1970), occurs on the Venezuelan mainland as the subspecies *A. p. venezuelae*. On Margarita and Los Frailes, on the other hand, it is represented by *A. p. margaritensis*. Further examples are: *Synallaxis albescens* represented by *S. a. occipitalis* in the unforested biotope of the mainland but by *S. a. nesiotis* in Margarita: *Xiphorhynchus picus* represented by *X. p. phalare* in the unforested biotope of the mainland but by *X. p. longirostris* on Margarita: and *Icterus nigrogularis* represented by *I. n. trinitatis* in the unforested biotope of the mainland and isolated savannas in Trinidad but by *I. n. helioeides* on Margarita.

I shall now list the Venezuelan islands inhabited by reptiles and amphibia. In this list the islands are put into three groups for purposes of comparison i.e. within the 200 m. isobath, between the 200 and 500 m. isobath, and outside the 1000 m. isobath. Margarita is not considered further. Local endemics are indicated by (e). Species and subspecies which only occur outside the 200 m. isobath are placed in brackets.

Islands within the 200 m. isobath

Aruba:
1. *Gonatodes albogularis albogularis*
2. *Gonatodes vittatus vittatus*
3. *Phyllodactylus julieni* (e)
4. *Thecadactylus rapicaudus*
5. *Anolis lineatus* [& Curaçao = (e)]
6. *Iguana iguana iguana*
7. *Ameiva bifrontata bifrontata*
8. *Cnemidophorus lemniscatus arubensis* (e)
9. *Gymnophthalmus laevicaudus*
10. *Tretioscincus bifasciatus*
11. *Leptodeira annulata*
12. *Crotalus durissus unicolor* (e)

Gentinela: nil.

Tortuga:
1. *Iguana iguana iguana*
2. *Cnemidophorus lemniscatus nigricolor*
3. *Phyllodactylus rutteni*

Cubagua:
1. *Gonatodes vittatus*
2. *Ameiva bifrontata bifrontata*
3. *Cnemidophorus lemniscatus*
4. *Phimophis guianensis*

Los Testigos:
1. *Gonatodes vittatus*
2. *Thecadactylus rapicaudus*
3. *Iguana iguana iguana*
4. *Tropidurus torquatus hispidus*
5. *Ameiva bifrontata bifrontata* (cf. RUTHVEN 1923, 1924)
6. *Dryadophis pleei*
7. *Oxybelis aeneus aeneus*

La Sola: nil.

Los Frailes:
1. *Gonatodes vittatus*
2. *Iguana iguana iguana*
3. *Tropidurus torquatus hispidus*
4. *Ameiva bifrontata bifrontata*
5. *Cnemidophorus lemniscatus*

Isla de Caribes:
1. *Phyllodactylus muelleri*

Coche:
1. *Gonatodes vittatus*
2. *Tropidurus torquatus hispidus*
3. *Cnemidophorus lemniscatus lemniscatus*

Islands within the 500 m. isobath but outside the 200 m. isobath

Blanquilla:
1. *Phyllodactylus rutteni*
2. (*Anolis bonariensis blanquillanus*)
3. *Iguana iguana iguana*
4. *Cnemidophorus lemniscatus nigricolor*

59

Los Hermanos: 1. *Phyllodactylus rutteni*
2. (*Anolis bonariensis blanquillanus*)
3. *Iguana iguana iguana*
4. *Cnemidophorus lemniscatus nigricolor*
5. *Tretioscincus bifasciatus*

Islands outside the 1000 m. isobath

Curaçao: 1. *Gonatodes albogularis albogularis*
2. (*Gymnodactylus antillensis*)
3. (*Phyllodactylus martini*)
4. *Thecadactylus rapicaudus* (cf. ROOY 1922)
5. (*Anolis lineatus*)
6. *Iguana iguana iguana*
7. (*Cnemidophorus murinus murinus*)
8. (*Gymnophthalmus lineatus*)
9. (*Dromicus antillensis*)
10. (*Leimadophis triscalis*)

Bonaire: 1. (*Gymnodactylus antillensis*)
2. (*Phyllodactylus martini*)
3. *Thecadactylus rapicaudus*
4. (*Anolis bonariensis bonariensis*)
5. *Iguana iguana iguana*
6. (*Cnemidophorus murinus ruthveni* (e))
7. (*Gymnophthalmus lineatus*)
8. *Leptotyphlops albifrons*

Los Roques: 1. (*Phyllodactylus rutteni*)
2. *Iguana iguana iguana*
3. *Cnemidophorus lemniscatus nigricolor*
4. (*Gonatodes vittatus roquensis* (e))

Orchila: 1. (*Phyllodactylus rutteni*)
2. (*Gymnodactylus antillensis*)
3. *Iguana iguana iguana*
4. *Cnemidophorus lemniscatus nigricolor*

Las Aves: 1. (*Gymnodactylus antillensis*)
2. *Iguana iguana iguana*
3. *Tropidurus torquatus hispidus*
4. *Cnemidophorus lemniscatus nigricolor*
5. *Gymnophthalmus laevicaudus*

Little Curaçao: 1. (*Gymnodactylus antillensis*)
2. (*Cnemidophorus murinus murinus* (e))

Little Bonaire: 1. (*Gymnodactylus antillensis*)
2. (*Phyllodactylus martinii* (e))
3. *Anolis bonariensis bonariensis*
4. *Cnemidophorus murinus ruthveni* (cf. BURT 1935)

The list makes clear that the proportion of endemics increases considerably outside the 500 m. isobath. It also shows that the distribution of many taxa most probably depends on passive drifting. This is also true, on the basis of studies so far made, for the molluscs (LORIE 1887, VERNHOUT 1914, BAKER 1924, 1925, GOULD 1969), birds (HARTERT 1893, CORY 1909, RUTTEN 1932) and odonata (GEIJSKES 1934). It is not certain that the big leguan (*Iguana iguana*) has spread naturally outside the 200 m. isobath. The same can be said for the gecko *Thecadactylus rapicaudus* which can very easily be introduced, like *Hemidactylus mabouia* (cf. MÜLLER 1969). This only indicates that a single case has little real significance (cf. *Crotalus durissus unicolor*, Barranquilla centre). For sweepstake dispersal by natural means can only be invoked if introduction by man can be excluded because of subspecific differentiation. The results show, however, that the islands near the mainland within the 200 m. isobath were populated from the Caribbean centre.

The great majority of the faunal elements of the Caribbean centre, including 97 subspecifically differentiated populations of birds, have their closest relatives in the Barranquilla, Roraima, Campo Cerrado and Caatinga centres.

20. The Roraima centre

This centre (fig. 5, no. 20) can be defined on the ranges of *Spinus magellanicus longirostris, Emberizoides herbicola duidae, Caprimulgus longirostris roraimae, Colibri coruscans germanus, Idioptilon margaritaceiventer auyantepui, Aratinga*

Fig. 27. Distribution of the polycentric and polytypic species *Spinus magellanicus* and *Caprimulgus longirostris*. *S. magellanicus longirostris* and *C. longirostris roraimae* are faunal elements of the Roraima centre.

pertinax chrysophrys and *Crotalus durissus ruruima* (fig. 27). By contrast with the montane forest fauna, which is very rich in endemics in the Roraima region (cf. Pantepui centre), the vertebrate fauna of the open landscape mostly consists of faunal elements distinct only at the subspecific level. An exception is the cricetine rodent *Podoxomys roraimae*.

The centre lies at an altitude of 600 to 1600 m. and includes the isolated savannas of Roraima, Uei-tepui, Cerro-Cuquenan, Gran Sabana, Mt. Twekquay, Aparaman-tepui, Auyan-tepui, Uaipan-tepui, Aprada-tepui, Chimantatepui, Acopan-tepui, Upuigma-tepui and Parai-tepui. The centre only applies to the savanna fauna and is therefore not comparable with the 'Pantepui' region studied by MAYR & PHELPS (1955, 1967) (cf. Pantepui centre).

The region of the Roraima centre has been well worked, especially as regards the birds and mammals (SCHOMBURG 1840, 1848, TATE 1928, 1930, 1932, TATE & HITCHCOCK 1930, CHAPMAN 1929, 1931, WHITELY 1884, PHELPS 1938, PHELPS & PHELPS 1962, AGUERREVERE, LOPEZ, DELGADO & FREEMAN 1939, ZIMMER & PHELPS 1944, 1946, 1948, 1949, 1952, GILLIARD 1941, PEBERDY 1941, MAYR & PHELPS 1955, 1967, WETMORE & PHELPS 1956). In addition the individual biotopes are already known to some extent from the floristic viewpoint (BROWN 1901, GLEASON 1929, EWAN 1950, MAGUIRE 1960, ARISTEGUIETA 1962, BUNTING 1963, STEYERMARK 1962, 1966, MAGUIRE 1965, ROBINSON 1965, SCHWEINFURTH 1967, SMITH 1967). Nevertheless no precise data yet exist as to the ecological valency of the vertebrates of the Roraima region.

The existence of endemic subspecifically distinct populations of typical indicator species of the South American open landscape shows that the isolated upland savannas of this area are natural. The fact that *Crotalus durissus* occurs here as a subspecifically distinct population certainly indicates that the centre has existed since the post-glacial arid phase (cf. discussion).

For some species adapted to an open landscape these upland savannas have served as a thoroughfare during a dry phase. This can be deduced from the degree of relationship of the *Crotalus durissus* populations in the high savanna on the Brazil-Guyana border and in the flood savanna of the Amazon basin (MÜLLER 1968, 1970).

The precise limits of the centre are not yet certain. This is because the degree of genetic relationship of the populations of the Roraima area, strongly isolated by rain forest, with those of the Gran Sabana has not yet been sufficiently worked out. OTREMBA (1954) has described the detailed patchwork of rain forests and savannas in this area.

21. The Pantepui centre

The centre (fig. 5, no. 21) can be defined on the ranges of several snakes i.e.: *Atractus duidensis, A. riveroi, A. steyermarki, Liophis ingeri, L. trebbaui, Thamnodynastes chimanta* and *Bothrops lichenosus*; on the range of the Leptodactylid *Eleutherodactylus marmoratus*; and on the ranges of 77 specifically or subspecifically distinct birds i.e. 86.5% of all bird species present with 2 monotypic genera, 23 endemic species and 52 endemic subspecies. This is based on

Fig. 28. The affinities of the 77 faunal elements of the bird fauna of the Pantepui centre. Thirty-one have their closest relatives in the Amazon centre and 46 in the montane forest centres.

64

data in CHAPMAN (1925, 1929, 1931), ROZE (1958, 1961, 1966) and MAYR & PHELPS (1967).

The prefered biotopes of the faunal elements of the centre (fig. 28) are the montane forests between Roraima (2,810 m.) in the east, Cerro Guaiquinima (1,800 m.) and Guanay (2,300 m.) in the north (Venezuela), Cerro Duida in the west, though only on the southern slopes, and Cerro de la Neblina (3,045 m.) in the south, on the border between Brazil and Venezuela.

Vertebrates other than birds have not been completely studied. Such data as exist, however, indicate a similar richness in endemics (BURT & BURT 1933, CABRERA 1957–1961, ROZE 1966). Thus for the mammals, all forms yet found in the centre must be considered as endemics, including the Cricetids *Podoxymys roraimae* and *Rhipidomys macconnelli*. Among marsupials, which have been well worked, the same is true for *Monodelphis brevicauda orinoci*, *Marmosa cinerea areniticola*, *M. murina roraimae*, *M. tyleriana phelpsi* and *M. tyleriana tyleriana*. The Teiids that occur in the centre are also endemic i.e. *Arthrosaura versteegii*, *Euspondylus leucostictus*, *Neusticurus tatei* and *Neusticurus rudis*; the closest relatives of these Teiids are found in the montane forest biomes of Colombia and Venezuela.

CHAPMAN (1931) and TATE (1932) believed that the high proportion of endemics must result from a situation where an originally continuous plateau connected the individual massifs. Thus CHAPMAN said (1931, p. 56): 'this fauna once occupied a much wider area of which these mountains formed a part ... remnants of once continuous table land which became dissected by erosion and is now represented by isolated fragments'. And TATE (1938, p. 472) wrote that: 'The development of the highly differentiated and endemic species may be held to have taken place in situ—by successive stages of adaptation to the environmental changes induced by the slowly rising plateau. Since the dissection of the area and its reduction to numerous faunal islands little marked change has taken place in the now geographically separated species. The presence of a species on one mountain top and its absence from another suggest only that some inimical condition has arisen on the latter to cause the species to die out'. This hypothesis can certainly be supported on geological grounds (GANSSER 1954, BEDERKE & WUNDERLICH 1968). On zoogeographical grounds, however, it must be rejected, since the taxa started to differentiate much too late. As the following table shows, there are different degrees of differentiation in the individual faunal elements among birds.

Endemic genera	2
Endemic species, closest related populations uncertain	4
Endemic species, closest related populations known	19
Endemic species, forming superspecies complexes with the closest related species	3
Non-endemic species with endemic subspecies	49
Non-endemic species without endemic subspecies	12
Total	89

The high proportion of endemic, subspecifically distinct populations shows that the differentiation of the Pantepui avifauna, in at least 49 cases, is most unlikely to have started in pre-glacial times (cf. Discussion). This indicates that differentiation began at a time when, in the opinion of geologists, a continuous plateau no longer existed. The presence of birds in montane forests that are often strongly isolated from each other can be explained in terms of their species-specific dispersion rates. Nevertheless 16 Pantepui faunal elements are represented by subspecifically distinct populations in individual montane forest biomes. This indicates that gene exchange is either limited or impossible.

The following species are represented by several subspecies in the Pantepui centre:

1. *Amazilia viridigaster cupreicauda & duidae.*
2. *Aulacorhynchus derbianus duidae & whitelianus & osgoodi.*
3. *Piculus rubiginosus guianae & paraquensis & viridissimus.*
4. *Xiphocolaptes promeropirhynchus tenebrosus & neblinae.*
5. *Synallaxis moesta macconnelli & griseipectus & yavii.*
6. *Automolus roraimae roraimae & duidae & paraquensis.*
7. *Lochmias nematura castanonota & chimantae.*
8. *Dysithamnus mentalis spodionotus & ptaritepui.*
9. *Myrmotherula behni inornata & yavii & camanii.*
10. *Percnostola leucostigma saturata & obscura.*
11. *Chamaeza campanisona yavii & obscura & fulvescens.*
12. *Knipolegus poecilurus salvini & paraquensis.*
13. *Platyrhynchus mystaceus ptaritepui & duidae & ventralis.*
14. *Mecocerculus leucophrys roraimae & parui.*
15. *Oxyruncus cristatus hypoglaucus & phelpsi.*
16. *Turdus olivater kemptoni & duidae & roraimae.*

An analysis of the origins of the 77 endemic birds shows that 31 have their closest relatives in the Amazon centre (cf. fig. 28), 21 in the montane forest of the Venezuelan coast, 5 in the North Colombian montane forest and 20 in the East Colombian montane forest. This last group shows close relationships in 5 cases to the isolated montane forest biomes in the Serra do Mar. Data at present available for the mammals give a similar picture. The snakes that must be taken as faunal elements of the centre either have connections with the montane forest biome of the Venezuelan coast (*Atractus*); or else have their closest relatives in the Amazonian lowland (*Thamnodynastes, Liophis, Bothrops*).

Fig. 29. Chamaeza campanisona. This is a typical montane forest species which is represented by a monocentric subspecies in the Pantepui centre.

Fig. 30. The ranges of the montane forest species *Aulacorhynchus derbianus* and *Myadestes leucogenys*, illustrating the close affinities between the Pantepui and Yungas centres.

22. The Guyanan centre

This centre (fig. 5, no. 22) extends over the Guyanan states, except for montane forest regions and savanna, and into south-east Venezuela (Amacupo, northern Bolivar). It is characterised by strong orographic subdivision and large areas of isolated high campos on the Brazilian border. In the north it is limited by a narrow strip of savanna along the Atlantic coast. In the south it is bounded, in order from east to west, by the high campos of Serra Tumuc-Humac, Avarai, Mapuera, Pacaraima including Roraima and Tapirapeco. This southern boundary is not at all so sharp for the faunal elements of the centre as the narrow strip of campo in the north (cf. fig. 31). For many Guyanan faunal elements

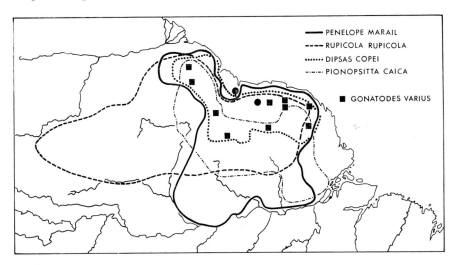

Fig. 31. Ranges of five faunal elements of the Guyana centre.

penetrate south and south-westwards between the higher mountains and reach their southern or western limits on the north bank of the Amazon or the east bank of the Rio Negro (fig. 31, fig. 32). It is remarkable that no centres of range of rain-forest species can be found in the immediate proximity of the high campos. Together with the endemic fauna as discussed under the Roraima centre this indicates that the high campos are natural rather than artificial.

The delta of the Orinoco together with the two islands of Trinidad and Tobago occupy a special position. Nevertheless they are inhabited mainly by Guyanan faunal elements. There are 37 species of snake in Trinidad and 20 in Tobago; of these, the percentage that came in via the Venezuelan coastal forest centre is roughly equal to the percentage that came from the Guyanan centre. Tobago lacks poisonous snakes (MERTENS 1969, 1972) but the only two Elapids of Trinidad i.e. *Micrurus circinalis* and *M. lemniscatus diutius* occur, apart from the island itself,

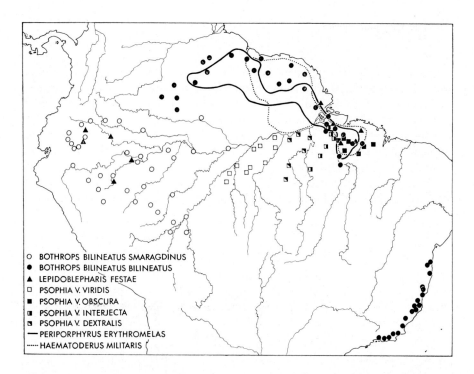

Fig. 32. Ranges of some rain-forest species in South America. The distribution of *Bothrops bilineatus* and of *Haematoderus militaris*, indicate the close affinity between the Guyanan and Para centres. Note the occurrence of an isolated *B. bilineatus* population in the Serra do Mar centre. This is closer related to the Guyanan populations of the species than to *B. b. smaragdinus* which occurs in the Amazon centre. The range of *Lepidoblepharis festae* is split in two by REINKE's savanna corridor, like the range of *Bolitoglossa altamazonica* (cf. fig. 19). The range of *Psophia viridis* and its subspecies illustrates the isolating effect of the rivers Tocantins, Xingu, Tapajoz and Madeira on populations of rain-forest birds.

in the Venezuelan coastal forest that extends west of Trinidad. On the other hand other widespread species, like *Lachesis mutus mutus*, *Tripanurgos compressus* and *Chironius carinatus* extend in from the Guyanas but cannot be reckoned as monocentric faunal elements. The influence of the Antilles on the faunas of the two islands is much smaller by contrast. Comparative figures, extracted from the work of JUNGE & MEES (1961), show convincingly that the two islands are closely related to the South American mainland (Table p. 71). Among the poikilothermous reptiles a similar comparison indicates an even more preponderant South American origin. At least the 37 snakes of Trinidad and the 20 of Tobago originated in the mainland opposite and show no connections with the Antilles. This statement is based on data from BOETTGER (1895), BARBOUR (1916), BRONGERSMA (1956), EMSLEY (1966), VERTEUIL (1968) and MERTENS (1969). The proportion of endemics on both islands is small (*Erythrolamprus aesculapii ocellatus*, *Ameiva ameiva tobagana* etc.). This like-

	Trinidad	Tobago
Migrant birds from the north	56 = 16%	36 = 25%
Migrant birds from the south	5 = 1.5%	1 = 0.7%
Endemic South American species	208 = 60%	6 = 41.4%
South America and Antilles	26 = 7.5%	18 = 12.5%
North and South America including Antilles	31 = 9%	18 = 12.5%
North America and Antilles	4 = 1%	7 = 4.9%
North and South America apart from Antilles	14 = 4%	3 = 2%
Endemic species	1 = 0.33%	0 = 0
Uncertain species	2 = 0.66%	0 = 0
	347 = 100%	144 = 100%

wise suggests that Trinidad and Tobago have not long been isolated. (cf. KENNY 1969). The existence of *Anolis richardii richardii* on Tobago points to a relationship with Grenada and St. Vincent. As a unique case, however, this has little importance and can be explained by passive drifting (cf. discussion under Caribbean centre).

There is an obvious barrier-zone for mainland species of South American origin north of Trinidad and Tobago. Despite this a few species, because of their vagility, have been able to pass beyond these islands in the direction of the Antilles (fig. 33). Some of these are very early immigrants, which mostly differ

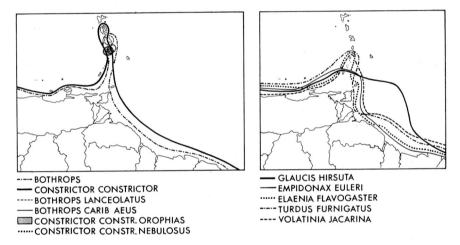

--- BOTHROPS
— CONSTRICTOR CONSTRICTOR
---- BOTHROPS LANCEOLATUS
— BOTHROPS CARIB AEUS
▨ CONSTRICTOR CONSTR. OROPHIAS
······ CONSTRICTOR CONSTR. NEBULOSUS

— GLAUCIS HIRSUTA
— EMPIDONAX EULERI
······ ELAENIA FLAVOGASTER
---- TURDUS FURNIGATUS
---- VOLATINIA JACARINA

Fig. 33. The northern limits of South American taxa in the region of Trinidad and Tobago.

markedly from their South American antecedents; others represent a stratum of relatively young arrivals. Among flying forms such as the bats and birds it cannot be excluded that continuous gene exchange takes place between the populations of the islands and the mainland.

The mammalian fauna presents much the same picture. HERSHKOVITZ (1969) derives most Trinidad species from the Venezuelan coastal forest centre. He nevertheless shows that they also have very close connections with the Guyanas.

A small percentage of species can be ascribed to the Caribbean centre, and occur in the isolated savannas of Trinidad. These cannot be used to support BEARD's supposition (1945) that during the Pleistocene, when the island was connected to the continent, the interior of Trinidad was extensively covered with savanna. The savannas scattered in the seasonal rain forest do show connections with the Venezuelan Llanos, both in their plants and animals. The differentiation of the populations occurring in them is nevertheless very small. Such populations probably reached Trinidad only in the post-glacial period. The fact that these connections are only found in birds and in flying insects support this view (cf. THOMPSON, 1963). The data of HOLLICK (1924) and BERRY (1925) on the Pleistocene flora of Trinidad are contradictory. From what HOLLICK (1924) says, however, it appears that the north of the island was already covered with forest in the Miocene.

Of all the bird species which occur in the Guyanan centre at least 29% can be reckoned as faunal elements of the centre (51 species, 139 subspecies). Among them are six monotypic genera: *Gypopsitta vulturina* (Psittacidae), *Haematoderus militaris* (Cotingidae), *Perissocephalus tricolor* (Cotingidae), *Microcochlearis josephinae* (Tyrannidae), *Cyanicterus cyanicterus* (Thraupidae), and *Lamprospiza melanoleuca* (Thraupidae).

The following bird species are Guyanan faunal elements: *Penelope marail, Crax alector, Pyrrhura egregia, Pionopsitta caica, Gypopsitta vulturina, Pionus fuscus, Neomorphus rufipennis, Caprimulgus maculesus, Threnetes niger, Phaethornis malaris, Anthracothorax viridigula, Lophornis ornata, Amazilia chionopectus, Galbula galbula, Monosa atra, Selenidera culik, S. nattereri, Ramphastos aurantiirostris, Picumnus spilogaster, Veniliornis sanguineus, Xiphorhynchus pardalotus, Synallaxis poliophrys* (but the systematic status of this species has not yet been satisfactorily worked out), *Myrmotherula guttata, M. gutturalis, Herpsilochmus stictocephalus, Gymnopithys rufigula, Cotinga cotinga, Iodopleura fusca, Rhytipterna immunda, Pachyramphus surinamus, Haematoderus militaris, Perissocephalus tricolor, Procnias alba, Phoenicircus carnifex, Rupicola rupicola, Pipra cornuta* (also known from two montane forest localities in the Roraima region), *Pipra serena, Corapipo gutturalis, Neopelma chrysocephalum, Phaeotriccus poecilocercus, Contopus albogularis, Todirostrum fumifrons, Microcochlearius josephinae, Phylloscartes virescens, Cyanocorax cayanus, Polioptila guianensis, Tanagra finschi, T. cayennensis, T. varia, Cyanicterus cyanicterus, Lamprospiza melanoleuca.*

The 656 breeding birds in the centre can be tabulated to show the number of species in each family. The number in front of the family name in the table is the number of species occurring. The number in brackets after the family name is the number of monocentric species. The table is compiled from data in PENARD & PENARD (1908–1910), REYNE (1921), HELLEBREKERS (1945), BLAKE (1961), MEYER DE SCHAUENSEE (1966), SNYDER (1966) and HAVERSCHMIDT (1968).

8 Tinamidae (0)
1 Podicipedidae (0)
0 Diomedeidae (0)

1 Procellariidae (0)
1 Hydrobatidae (0)
1 Pelecanidae (0)

o Phaethornitidae (o)
1 Pelecanidae (o)
1 Sulidae (o)
1 Phalocrocoracidae (o)
1 Anhingidae (o)
1 Fregatidae (o)
13 Ardeidae (o)
1 Cochlearidae (o)
3 Ciconiidae (o)
7 Threskiornithidae (o)
1 Phoenicopteridae (o)
1 Anhimidae (o)
13 Anatidae (o)
3 Cathartidae (o)
33 Accipitridae (o)
1 Pandionidae (o)
12 Falconidae (o)
6 Cracidae (2)

2 Phasianidae (o)
1 Opisthocomidae (o)
1 Aramidae (o)
1 Psophiidae (o)
16 Rallidae (o)
1 Heliornithidae (o)
1 Eurypygidae (o)
o Cariamidae (o)
1 Jacanidae (o)
o Rostratulidae (o)
o Haematopodidae (o)
5 Charadriidae (o)
11 Scolopacidae (o)
1 Recurvirostridae (o)
o Phalaropodidae (o)
1 Burhinidae (o)
o Thinocoridae (o)
o Chionidae (o)
1 Stercorariidae (o)
8 Laridae (o)
1 Rynchopodidae (o)
14 Columbidae (o)
31 Psittacidae (3)

11 Cuculidae (1)

1 Tytonidae (o)
12 Strigidae (o)
1 Steatornithidae (o)
4 Nyctibiidae (o)
13 Caprimulgidae (1)

9 Apodidae (o)
40 Trochilidae (6)

7 Trogonidae (o)
3 Alcedinidae (o)
1 Momotidae (o)
7 Galbulidae (1)

8 Bucconidae (1)

1 Capitonidae (o)
10 Ramphastidae (3)

23 Picidae (2)

20 Dendrocolaptidae (1)

25 Furnariidae (1)

52 Formicariidae (4)

2 Conopophagidae (o)
o Rhinocryptidae (o)
33 Cotingidae (8)

1 Rupicolidae (1)

15 Pipridae (4)

78 Tyrannidae (5)

1 Oxyruncidae (o)
o Phytotomidae (o)
o Alaudidae (o)
11 Hirundinidae (o)
3 Corvidae (1)

o Cinclidae (o)
10 Troglodytidae (o)
2 Mimidae (o)
12 Turdidae (o)
4 Sylviidae (1)

1 Motacillidae (o)
o Bombycillidae (o)
o Ploceidae (o)
11 Vireonidae (o)
21 Icteridae (o)
14 Parulidae (o)
11 Coerebidae (o)
1 Tersinidae (o)
40 Thraupidae (5)

o Catamblyrhynchidae (o)
33 Fringillidae (o)

It appears that only in the rain-forest family of the Rupicolidae are all the species that are present endemic. The other genera of the various families, except *Myrmotherula*, never have more than one species that must be reckoned as a faunal element of the centre.

The individual polycentric species that occur in the Guyanan centre can be analysed according to their subspecific differentiation. At least 139 polycentric and polytypic species have a monocentric subspecies range that coincides with the Guyanan centre (cf. fig. 35 among others).

The reptiles and amphibia, though less mobile, seem to have similar figures for endemic species to the birds. This result is somewhat uncertain, however, because the ecological valency of some reptile and amphibian species has not been well enough worked out for the purposes of our problem (cf. QUELCH 1899, PARKER 1935, SCHMIDT 1939, HOGE 1962, BRONGERSMA 1966).

In the families Pipidae and Leptodactylidae six species can be reckoned as faunal elements of the Guyanan centre. These are: *Pipa aspera, Ceratophrys cornuta, Eleutherodactylus beebei, E. grandoculis, E. inguinalis, E. lineatus.* Further Anuran faunal elements are: *Dendrobates tinctorius, D. azureus* and *Allophryne ruthveni.*

I have studied three Iguandis and four Teiids among the faunal elements of the centre and confirmed their validity. They are: *Anolis chrysolepis* (Iguanidae), *A. nitens* (Iguanidae), *Uranoscodon superciliosa* (Iguanidae), *Bachia schlegeli* (Teiidae), *Leposoma percarinatum* (Teiidae), *Neusticurus rudis* (Teiidae), *Arthrosaura versteegi* (Teiidae), *Amphisbaena stejnegeri* (Amphisbaenidae).

The relationships between the Guyanan and Para centres will be dealt with under the latter (cf. fig. 32).

There is a particularly close relationship between the Guyanan and Amazon centres. In many cases, nevertheless, the populations that occur in the two centres are subspecifically distinct, so that no difficulty arises in referring a population to its distribution centre (cf. the ranges of *Bothrops bilineatus bilineatus* and *B. b. smaragdinus* in fig. 32). Besides these forms, polycentric and monotypic species also occur in both centres. In such cases ascription to a centre is only possible if there is clinal variation e.g. *Galbula galbula* is a Guyanan faunal element.

Knipolegus orenocensis and *Phaeotriccus poecilocercus* indicate a close relationship between the Guyanan and Madeira centres.

23. The Para centre

The centre (fig. 5, no. 23) is limited in the west and north by the Rio Tocantins, and by the northern part of the Rio Araguaia south to Matheus; in the south it is limited by the Serra do Gurupi of the northern part of Maranhao and by the Rio Grajau, though some species extend as far as the Rio Parnaiba; in the east it is limited by the Rio Guaná and by the restinga of the Atlantic coast. *Pyrrhura perlata* and *Xipholena lamellipennis*, which are both Para faunal elements, reach their western limits only at the east bank of the Rio Xingu (fig. 34). Even so it can be shown that their dispersal centre lies in the nuclear area just defined.

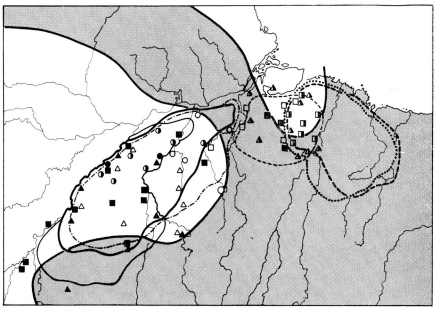

○ RHEGMATORHINA GYMNOPS	● PHLEGOPSIS BORBAE
--- ARATINGA GUAROUBA	— DENDROCOLAPTES CONCOLOR
△ PYRRHURA RHODOGASTER	▲ MYRMOTHERULA SCLATERI
■ CONOPOPHAGA MELANOGASTER	□ PIPRA IRIS
—— PIPRA NATTERERI	--- ODONTORCHILUS CINEREUS
▲ PYRRHURA PERLATA	◨ GYMNOSTINOPS BIFASCIATUS
····· ORTALIS SUPERCILIARIS	▬▬ CONOPOPHAGA ROBERTI
◑ PENELOPE PILEATA	○ AWI-KLIMATE

Fig. 34. Faunal elements of the Para and Madeira centres. Para faunal elements are: *Aratinga guarouba, Pyrrhura perlata, Ortalis superciliaris, Pipra iris, Gymnostinops bifasciatus* and *Conopophaga roberti*. Madeira faunal elements are: *Pyrrhura rhodogaster, Conopophaga melanogaster, Pipra nattereri, Penelope pileata, Phlegopsis borbae, Dendrocolaptes concolor, Myrmotherula sclateri, Odontorchilus cinereus*. Areas covered by savanna during the post-glacial arid phase are indicated by shading (AWI klimate after REINKE 1962).

The rain forest of the island of Marajó, which lies just north, has only a slight connection with the Para centre in its fauna. Thus only four Para elements have settled in the rain forest of Marajo i.e. *Pyrrhura perlata*, *Xipholena lamellipennis*, *Terenitriccus erythrurus hellmayri* and *Pipra chloris grisescens*. Other species that occur both in the Para centre and on Marajo are polycentric.

The same rather weak relationship appears in the herpetofauna. Some typical Para elements are represented both in the Para centre and on Marajo i.e. *Aulura anomala*, *Iphisa elegans*, *Leimadophis oligolepsis* and *Amphisbaena mitchelli*. The other species which occur in both, however, have very widespread and sometimes strongly disjunct distributions. The Para faunal elements *Pipa snethlagae*, *Eupemphix paraensis* and *Leptodactylus matinezi* have not been found on Marajo (cf. MÜLLER 1969).

Bothrops bilineatus (cf. fig. 32), which I found on Marajo in 1964, has a very interesting range. It shows connections on the one hand with the Para, Guyanan and Amazon centres and on the other with the Serra do Mar centre. At the same time the relationship is much closer between the Para, Guyanan and Serra do Mar populations (same subspecies) than between them and the Amazon centre (subspecies *B. b. smaragdinus*). It would be premature to conclude from this, however, that the Para centre and Serra do Mar centre are very closely related.

HELLMAYR (1912) mentioned a number of species of birds that occurred exclusively in the regions of the Para centre and the Serra do Mar centre. However, if his localities are examined carefully, it is obvious that none of these species are strictly adapted to rain forest. Thus of the species mentioned by him, I have never been able to observe *Mimus lividus* (= *M. saturninus*), *Stelgidopteryx ruficollis ruficollis* or *Thraupis palmarum palmarum* in thick rain forest, nor to catch them there with Japanese nets. However I saw them on the flood campo of the island of Marajo at Santa Cruz do Arari in January 1965 and March 1969, at São Jose dos Campos (state of São Paulo) in October 1964 and on the Rio Cuiaba (near Cuiaba, Mato Grosso) in the Campo Cerrado in February 1965.

Again there is a herpetological example which is supposed to show the relationship between the forests of the Serra do Mar and the Para centre (CUNHA 1961) but which seems dubious to me. This is the Anguid *Ophiodes striatus* which, as I have been able to show in several places in its south-east Brazilian range, is in no way a rain-forest form (MÜLLER 1968). Moreover its occurrence in Para is doubtful. Thus Cunha mentions the species from the state of Para but I have not been able to find any actual specimens—they are lacking for instance in the Museum Goeldi in Belem. An old specimen from Surinam in the Leiden Museum, without precise locality or the name of the collector, also seems problematical since this species cannot now be found in Surinam (personal communication, Mr. HOOGMOED, Leiden).

Thus among the herpetofauna only *Eleutherodactylus binotatus* can be used to show a relationship between the Serra do Mar and Para centres. For I have met with it as a typical rain-forest species in the Serra do Mar (MÜLLER 1968) and have also found it in the forests round Belem.

76

The numerous species distributed both in the Serra do Mar and Amazon centres are mostly polycentric and widespread in the Amazon area. Only in a few cases are they limited to one centre (cf. discussion of Serra do Mar centre).

In addition there are species with disjunct ranges which have been found up till now only in the Amazon and Para centres. This disjunction cannot yet be clearly interpreted, but is found in fishes (cf. GERY 1969) and reptiles (*Lepidoblepharis festae*, cf. fig. 32). It also holds for the mammals *Tamarin tamarin* and *Dasypus kappleri* though these two species also occur in the southern part of the Guyanan centre. The disjunction is certainly real for all the species mentioned and not due merely to ignorance of the true distribution. CUNHA (1961) indeed considers *L. festae* as a typical Amazon species found in: 'toda a planicie Amazonica, desde os contrafortes orientais dos Andes'. However actual specimens to support his view are completely lacking (VANZOLINI 1968). I have looked for this gecko in rain forest in Manaus, Uaupes and Santarem (MÜLLER, 1970) without finding it.

A few species of the Amazon centre, mostly belonging to the Ucayali sub-centre, extend along the southernmost Amazonian forests into the Para centre (*Crax mitu mitu, Phlegopis nigromaculata* etc. cf. fig. 35). In some cases they then become subspecifically distinct (*Phlegopsis nigromaculata paraensis*).

The affinities between the Para and Madeira centres are greater than those between the Para and Amazon centres. This is particularly obvious in the case of *Psophia viridis* whose species range includes both centres (fig. 32). This species range is made up of four subspecies ranges whose limits are determined by the river systems of the Madeira, Tapajoz, Xingu and Tocantins. Moreover the subspecies are not all equally related to each other, so that *Psophia viridis viridis*, found from the Madeira to the Tapajoz, is closest related to *P. v. dextralis*, found from the Tapajoz to the Xingu. On the other hand the populations east of the Xingu (*P. v. interjecta*) are more closely related to the Para form (*P. v. obscura*) than to *P. v. dextralis*.

If the Para centre is compared quantitatively with the Amazon, Guyanan and Madeira centres, it is found to have the smallest number of endemic species. Nevertheless at least 10% of the bird fauna of the Para centre must be considered as faunal elements of the centre. Most of these Para endemics belong to families that MAYR considers to be Neotropical in origin and in which most of the species are typical rain-forest birds e.g. Formicariidae, Cotingidae, Pipridae and Ramphastidae. They can be listed as follows:

Crax fasciolata pinima (Cracidae)
Ortalis superciliaris (Cracidae)
Pipile cujubi (Cracidae)
Psophia viridis obscura (Psophiidae)
Aratinga guarouba (Psittacidae)
Pyrrhura perlata (Psittacidae)
Deroptyus accipitrinus fuscifrons (Psittacidae)
Pionites leucogaster leucogaster (Psittacidae)
Phaethornis superciliosus muelleri (Trochilidae)
Campylopterus obscurus obscurus (Trochilidae)

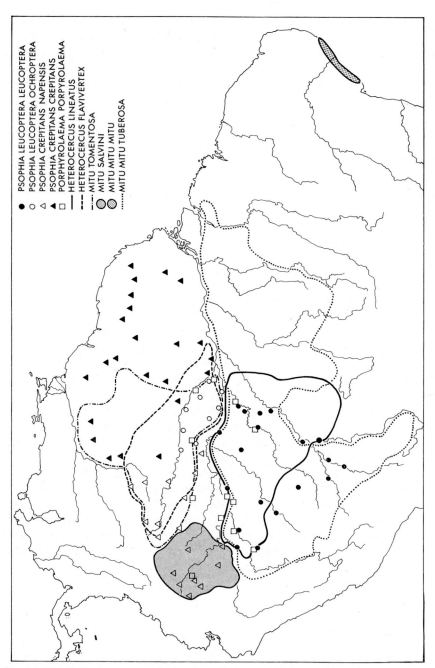

Fig. 35. Distribution types among Amazon, Guyanan and Serra do Mar faunal elements. *Porphyrolaema porphyrolaema* occurs both in the Napo and Ucuyali subcentres of the Amazon centre. On the other hand, *Psophia leucoptera leucoptera*, *Heterocercus lineatus* and *Mitu mitu tuberosa* of the Ucuyali subcentre are replaced by close relations in the Napo subcentre or the Serra do Mar subcentre, as the case may be. *Psophia crepitans crepitans* is a Guyanan faunal element. *P. crepitans napensis* is a faunal element of the Napo subcentre.

PSOPHIA LEUCOPTERA LEUCOPTERA
PSOPHIA LEUCOPTERA OCHROPTERA
PSOPHIA CREPITANS NAPENSIS
PSOPHIA CREPITANS CREPITANS
PORPHYROLAEMA PORPYROLAEMA
HETEROCERCUS LINEATUS
HETEROCERCUS FLAVIVERTEX
MITU TOMENTOSA
MITU SALVINI
MITU MITU MITU
MITU MITU TUBEROSA

Agyrtia nitidifrons (Trochilidae)
Thalurania furcata furcatoides (Trochilidae)
Heliothrix aurita phainolaema (Trochilidae)
Momotus momota paraensis (Momotidae)
Pteroglossus bitorquatus bitorquatus (Ramphastidae)
Selenidera gouldi (Ramphastidae)
Piculus chrysochloros paraensis (Picidae)
Picumnus minutissimus pallidus (Picidae)
Celeus undatus multifasciatus (Picidae)
Synallaxis rutilans omissa (Furnariidae)
Dendrexetastes rufigula paraensis (Furnariidae)
Thamnophilus aethiops incertus (Formicariidae)
Dysithamnus mentalis emiliae (Formicariidae)
Thamnomanes caesius hoffmannsi (Formicariidae)
Myrmotherula hauxwelli hellmayri (Formicariidae)
Pyriglena leuconota leuconota (Formicariidae)
Hypocnemis poecilinota vidua (Formicariidae)
Phelgopsis nigromaculata paraensis (Formicariidae)
Conopophaga roberti (Conopophagidae)
Xipholena lamellipennis (Cotingidae)
Pipra iris (Pipridae cf. HAFFER 1971, FRY 1970)
Pipra chloris grisescens (cf. NOVAES 1964)
Todirostrum illigeri (Tyrannidae)
Todirostrum illigeri (Tyrannidae)
Todirostrum sylvia schulzi (Tyrannidae)
Terenotriccus erythrurus hellmayeri (Tyrannidae)
Gymnostinops bifasciatus (Icteridae)
Granatellus pelzelni paraensis (Parulidae)
Tangara velia signata (Thraupidae).

The position and limits of the Para centre in the north, east and south can be explained well enough from present-day ecology. Stretches of water, even narrow ones, act as barriers to strict rain-forest species (cf. MÜLLER 1966, 1968, 1970) and open landscapes have a similar effect.

Endemic rain-forest species are lacking in the region between the Xingu and the Tocantins. Four subspecifically distinct bird populations in this region have their closest relationships with the Para centre (cf. *Psophia viridis*). If the limits of range of Para and Madeira faunal elements are compared, it is easy to show that this area between the Xingu and the Tocantins is a 'zone of impoverishment' for rain-forest species (see fig. 34). It is partly inhabited by species whose dispersal centres lie in the open landscapes of central Brazil.

24. The Madeira centre

This centre (fig. 5, no. 24) is limited in the north by the Amazon, in the west by the Madeira and Beni, in the east by the Xingu and in the south by the Bolivian Eastern Cordillera. All Madeira faunal elements are strictly limited by the west and north boundaries of the centre (fig. 35), but a few species penetrate beyond the southern and eastern boundaries, particularly along gallery forests.

Madeira faunal elements reach their southern limits in the Montana zone of the Bolivian Yungas into which they sometimes penetrate (NIETHAMMER 1953). They are lacking in the subtropical zone i.e. in the Medio Yungas (cf. CHAPMAN 1926, MEYER DE SCHAUENSEE 1964). The Medio Yungas, the cloud forests (Ceja) above them, and the paramos of the Bolivian Eastern Cordillera form a sharp faunal boundary (NIETHAMMER 1953).

The occurrence of Campo Cerrado faunal elements such as *Melanoporeia torquata* in the area of the Serra do Cachimbo and the Cururu region shows that the Madeira centre also has a limit in that place. The specimens that I have seen from this region confirm SIOLI's interpretation (1967) that a transition between rain forest and Campo Cerrado exists here.

Sixteen species of birds can be reckoned as faunal elements of the Madeira centre. These are:

Pteroglossus bitorquatus madeirensis
Dendrocolaptes hoffmannsi
Myrmotherula iheringi
Myrmeciza stictothorax
Rhegmatorhina gymnops
R. berlepschi
R. hoffmansi

Phlegopsis borbae
Grallaria berlepschi
Conopophaga melanogaster
Tityra leucura
Pipra nattereri
Todirostrum senex (the holotype from Borba on the east bank of the Madeira is the only specimen known),
Idioptilon aenigma
Odontorchilus cinereus
Pipile cujubi

The importance of the Madeira centre for the South American forest fauna is shown by the way in which the ranges of polycentric species are divided into subspecies ranges. Thus the widespread Tinamid *Tinamus tao* occurs between the Madeira and the Xingu as the easily diagnosed nominate form. A close relative, *Tinamus major* is represented in the same centre by the endemic subspecies *T. m. olivaceus*.

The way in which the range of *Tinamus major* is subdivided into subspecies ranges demonstrates the importance of the various forest centres of Amazonia. The nominate form is a faunal element of the Guyanan centre; the subspecies *T. m. olivaceus* is a faunal element of the Madeira centre and the two subspecies *T. m. serratus* and *T. m. peruvianus* belong to the Amazon centre. Within the Amazon centre the subspecies *T. m. serratus* inhabits the Napo subcentre and *T. m. peruvianus* the Ucayali subcentre.

According to the work of SNETHLAGE (1906–1907), PINTO (1932, 1938–1944, 1947, 1949, 1964) and SICK (1960) at least 53 bird populations are sub-specifically distinct in the Madeira centre.

For the mammals a similar high figure probably applies (CABRERA 1957–1961, HERSHKOVITZ 1969, AVILA-PIRES 1969). Thus two species of the genus *Callithrix* which has eight species in total (AVILA-PIRES 1969), must be reckoned as faunal elements of the Madeira centre. According to CABRERA (1957–1961) 11 species and 33 subspecies of mammals are faunal elements of the centre.

25. The Amazon centre

The Amazon centre (fig. 5, no. 25) is limited in the west by the montane forest biome of the Andes, in the north by the dry area of the Venezuelan Llanos, in the north-east by the Rio Negro and the Serra de Unturan, and in the south-east by the Rio Madeira (fig. 36). A few Amazon faunal elements occur along the south-east Andean rain forests right into the Guaporé region (*Crax mitu*).

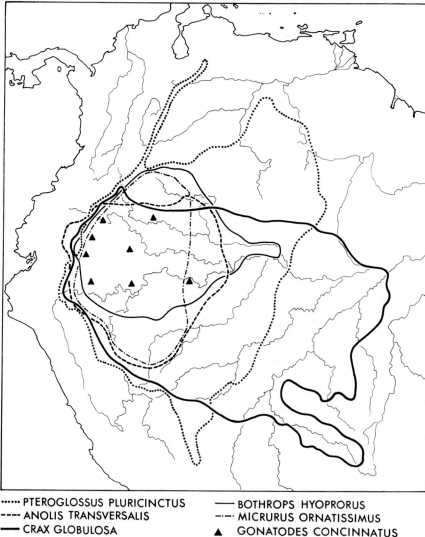

····· PTEROGLOSSUS PLURICINCTUS   —— BOTHROPS HYOPRORUS
---- ANOLIS TRANSVERSALIS   —·— MICRURUS ORNATISSIMUS
—— CRAX GLOBULOSA   ▲ GONATODES CONCINNATUS

Fig 36. Types of range among faunal elements of the Amazon centre. For *G. concinnatus* cf. Rivero Blanco 1968, Mem. soc. Ccien. Nat. La Salle Caracas 27: 104.

Others penetrate to the east, across the Madeira, as far as the Rio Xingu (e.g. *Myrmeciza hemimelaena, Attila bolivianus, Heterocercus lineatus*), the Rio Tocantins (*Picumnus aurifrons, Myrmotherula ornata*) or the Rio Tapajoz (*Liosceles thoracicus, Elaenia pelzelni*). This penetration depends partly on the range of action of the species named and partly, no doubt, on the fact that they meet with no competing species in the Madeira centre, which is relatively poor in species.

The species ranges of the endemic birds of the Amazon centre can often be subdivided into two subspecies ranges (cf. HAFFER 1969, MÜLLER 1972). The Amazon river, including the Solimões, and the Rio Marañon, act as a barrier to many rain-forest species, and the populations so separated form a northern and a southern subspecies. The northern subspecies, belonging to the Napo subcentre, have the Rio Negro as their north-east limit and the Amazon as their southern limit. The southern subspecies, belonging to the **Ucayali subcentre,** have the Amazon as their northern limit and the Rio Madeira as their south-east limit. A number of species, as opposed to subspecies, are also restricted in range to the southern or Ucayali subcentre. Examples are:

Crypturellus bartletti	*Galbula cyanescens*
C. strigulosus	*G. pastazae*
Odontophorus kuhli	*Malacoptila semicincta*
Aratinga weddellii	*Nonnula sclateri*
Phaethornis philippi	*Rhematorhina melanostica*
Leucippus chlorocercus	*Heterocercus lineatus*
Galbalcyrhynchus leucotis	*Muscisaxicola fluviatilis*
Brachygalba albogularis	

On the other hand there are fewer species that are limited to the northern or **Napo subcentre** (cf. HAFFER). Examples are: *Nonnula amaurocephala* and *Heorcercus flavivertex*. This unequal distribution of species ranges, with 67% in both subcentres, 28% in the Ucayali subcentre and 5% in the Napo subcentre is, however, only obvious in the birds. In reptiles and amphibia it is less clear. This must be due partly to the vagility of these groups, since water barriers are not important for amphibians, especially their larval stages. It also partly reflects the slower rate of differentiation in reptiles and amphibians. It is interesting to compare in this connection the differentiation of amphibians, reptiles, birds and mammals on the island of São Sebastião (MÜLLER 1966, 1968).

Together with the Guyanan centre, the Amazon centre has been of the greatest importance for the evolution of the South American forest fauna. Not less than 157 species of birds can be reckoned as faunal elements of this centre. They are:

Tinamus guttatus	*Crax annulata*
Crypturellus bartletti	*C. globulosa*
C. strigulosus	*C. urumutum*
Odontophorus stellatus	*Mitu salvini*
Leucopternis kuhli	*Nothocrax urumutum*
L. schistacea	*Psophia leucoptera*
Micrastur buckleyi	*Anurolimnas castaneiceps*

Aramides calopterus
Ara couloni
Aratinga weddellii
Pionopsitta barrabandi
Pyrrhura albipectus
P. calliptera
Brotogeris cyanoptera
Pionites leucogaster
Neomorphus pucheranii
Pulsatrix melanota
Nyctibius bracteatus
Chordeiles rupestris
Phaetornis philippi
P. hispidus
Eutoxeres condamini
Phlogophilus hemileucus
Leucippus chlorocercus
Heliodoxa schreibersii
H. gularis
Galbalcyrhynchus leucotis
Brachygalba albogularis
Galbula tombacea
G. cyanescens
G. pastazae
Malacoptila semicincta
M. rufa
Micromonacha lanceolata
Nonnula sclateri
N. amaurocephala
N. brunnea
Monasa flavirostris
Capito aurovirens
Eubucco richardsoni
Pteroglossus mariae
P. olallae
P. beauharnaesii
P. flavirostris
Selenidera reinwardtii
Andigena cucullata
Ramphastos cuvieri
R. culminatus
Picumnus rufiventris
P. borbae
P. castelnau
P. pumilus
P. aurifrons
Dendrexetastes rufigula
Hylexetastes stresemanni
Xiphorhynchus elegans
X. ocellatus
Furnarius minor
Synallaxis albigularis
S. cherriei
S. moesta
Certhiaxis mustelina

Automolus melanopezus
Philydor erythropterus
Ancistrops strigilatus
Automolus dorsalis
Metopothrix aurantiacus
Frederickena unduligera
Tripadectes holosticus
T. melanorhynchus
Megasticus margaritatus
Neoctantes niger
Myrmotherula obscura
M. cherriei
M. ambigua
M. hauxwelli
M. guttata
M. assimilis
M. ornata
M. leucophthalma
M. erythrura
M. haematonota
Thamnophilus praecox
Thamnomanes saturnius
T. schistogynus
Dichrozona cincta
Herpsilochmus dorsimaculatus
Myrmochanes hemileucus
Drymophila devillei
Terenura humeralis
Myrmoborus lugubris
M. myotherinus
M. melanurus
Cercomacra serva
Percnostola schistacea
Hypocnemis hypoxantha
Myrmeciza pelzelni
M. hemimelaena
M. hyperythra
M. goeldi
M. melanoceps
M. fortis
Gymnopithys salvini
Rhegmatorhina melanosticta
R. cristata
Phlegopsis erythroptera
Chamaeza nobilis
Formicarius ruficrons
Hylophylax punctulata
Liosceles thoracicus
Thamnocharis dignissima
Porphyrolaema porphyrolaema
Grallaria berlepschi
G. przewalskii
Conopophaga peruviana
Scytalopus macropus
Cotinga maynana

84

Conioptilon mcilhennyi
Xipholena punicea
Pipreola chlorolepidota
Attila bolivianus
A. citriniventris
Lipaugus subalaris
Heterocercus lineatus
H. flavivertex
Muscisaxicola fluviatilis
Todirostrum poliocephalum
T. capitale
Myiophobus cryptoxanthus
Ramphotrigon fuscicauda
Idioptilon rufigulare
Lophotriccus eulophotes
Serpophaga hypoleuca
Elaenia gigas

E. pelzelni
Thryothorus griseus
Turdus lawrencii
Hylophilus olivaceus
Ocyalus latirostris
Clypicterus oseryi
Cacicus sclateri
Cyanerpes nitidus
Conirostrum margaritae
Tangara callophrys
T. schrankii
Ramphocelus nigrogularis
R. melanogaster
Tachyphonus rufiventer
Lanio versicolor
Caryothraustes humeralis

Of these species, 14 belong to monotypic genera. These are:

Anurolimnas castaneiceps (Rallidae)
Micromonacha lanceolata (Bucconidae)
Metopothrix aurantiacus (Furnariidae)
Ancistrops strigilatus (Furnariidae)
Megastictus margaritatus (Formicariidae)
Neoctantes niger (Formicariidae)
Myrmochanes hemileucus (Formicariidae)
Dichrozona cincta (Formicariidae)
Porphyrolaema porphyrolaema (Formicariidae)
Liosceles thoracicus (Rhinocryptidae)
Conioptilon mcilhennyi (Cotingidae)
Ocyalus latirostris (Icteridae)
Clypicterus oseryi (Icteridae)
Dendrexetastes rufigula (Dendrocolaptidae).

In addition there are 173 polycentric species with subspecies that must be reckoned as faunal elements of the centre. In all, there are therefore 330 faunal elements of the Amazon centre in the avifauna (fig. 37).

The number of Amazon faunal elements in the amphibia must be equally high. Even within the two well worked families of the Pipidae and Leptodactylidae there are 48 species which must be reckoned as faunal elements. These are:

Pipa pipa
Ceratophrys calcarata
C. testudo
Edalorhina perezi
Eleutherodactylus acuminatus
E. altamazonicus
E. appendiculatus
E. brachypodius
E. brevicrus
E. buckleyi

E. bufonius
E. calcaratus
E. carvalhoi
E. conspicillatus
E. curtipes
E. galdi
E. glandulosus
E. granulosus
E. koki
E. leptodactyloides

85

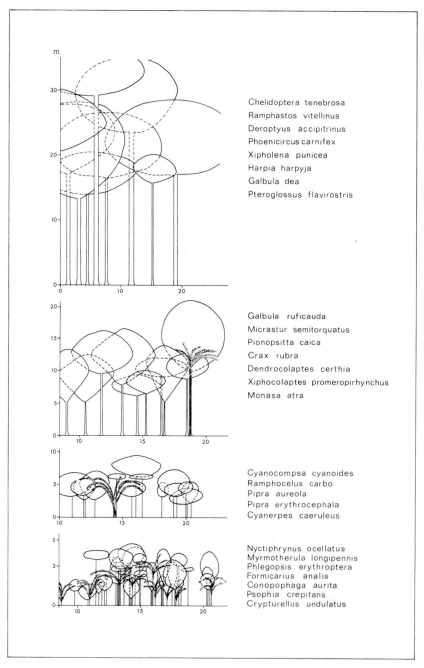

Chelidoptera tenebrosa
Ramphastos vitellinus
Deroptyus accipitrinus
Phoenicircus carnifex
Xipholena punicea
Harpia harpyja
Galbula dea
Pteroglossus flavirostris

Galbula ruficauda
Micrastur semitorquatus
Pionopsitta caica
Crax rubra
Dendrocolaptes certhia
Xiphocolaptes promeropirhynchus
Monasa atra

Cyanocompsa cyanoides
Ramphocelus carbo
Pipra aureola
Pipra erythrocephala
Cyanerpes caeruleus

Nyctiphrynus ocellatus
Myrmotherula longipennis
Phlegopsis erythroptera
Formicarius analis
Conopophaga aurita
Psophia crepitans
Crypturellus undulatus

Fig. 37. Vertical distribution of 27 forest species (Aves) in the Manaus area. The forest stratification is after FITTKAU & KLINGE. The horizontal axis, like the vertical, is graduated in metres.

86

E. nigrovittatus
E. palmeri
E. pastazensis
E. platydactylus
E. rosmelinus
E. sulcatus
E. tachyblepharis
E. ventrimarmoratus
E. ventrivittatus
E. vilarsi
Eupemphix petersi
E. scherei
Hydrolaetare schmidti (monotypic genus)
Leptodactylus dantasi

L. discodactylus
L. hemidactyloides
L. hololius
L. intermedius
L. macroblepharus
L. melini
L. nigrescens
L. rhodonotus
L. rhodostigma
L. romani
L. rubido
L. tuberculosus
L. vilarsi
Syrrhophus chalceus

13 Iguanid species and 18 Teiid species also appear to be Amazon faunal elements if the locality data are considered. These are:

Iguanidae
Anolis punctatus boulengeri
Anolis transversalis
Anolis bombiceps
Anolis laevis
Anolis leptoscelis
Anolis scapularis
Ophryoessoides aculeatus
Ophryoessoides scapularis
Polychrus liogaster
Enyalioides laticeps
Enyalioides oshaughnessyi
Enyalioides praestabilis
Uracentron guentheri

Teiidae
Ophiognomon abendrothii
Ophiognomon trisanale
Ophiognomon vermiformis
Ptychoglossus brevifrontalis
Ptychoglossus picticeps
Kentropyx altamazonicus
Kentropyx pelviceps
Leposoma parietale
Morunasaurus annularis
Neusticurus cochranae
Neusticurus ecpleopus
Neusticurus strangulatus
Prionodactylus argulus
Prionodactylus manicatus
Proctoporus bolivianus
Arthrosaura reticula
Alopoglossus copii
Alopoglossus buckleyi

The Amazon centre could probably be divided not merely into subcentres but at a tertiary or even quaternary level. Thus in the Ucayali subcentre certain taxa seem to have geographical ranges of a special type in the areas between the rivers Jurua and Beni and the Andean highlands. However, this region is still too badly known from the distributional point of view to take this as indicating an independent dispersal centre (fig. 38, cf. ranges of *Gonatodes hasemanni*, *Ramphocelus carbo connectens*).

In the Napo subcentre the influence of the Guyanan centre is stronger than in the Ucayali subcentre. Guyanan faunal elements such as *Galbula galbula* reach the north bank of the Amazon-Solimões river. These forms are certainly exceptional however. The Rio Negro is an obvious boundary for most Guyanan faunal elements as it is for most Amazon faunal elements in the opposite direction.

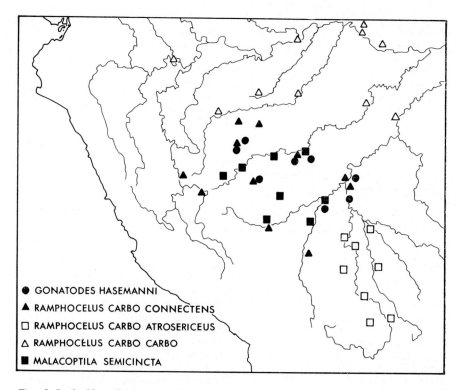

GONATODES HASEMANNI
▲ RAMPHOCELUS CARBO CONNECTENS
☐ RAMPHOCELUS CARBO ATROSERICEUS
△ RAMPHOCELUS CARBO CARBO
■ MALACOPTILA SEMICINCTA

Fig. 38. In the Ucayali subcentre of the Amazon centre there is a type of geographical range, stretching from the Jurua and Beni on the one hand to the Andean highlands on the other, which suggests that an independent dispersal centre may exist in this region.

88

26. The Yungas centre

This centre (fig. 5, no. 26, fig. 39) is indicated by the ranges of *Mazama chungi* (Cervidae), *Atlapetes fulviceps* (Fringillidae), *Hemispingus trifasciatus* (Thraupidae), *Marmosa yungasensis* (Didelphidae), *M. aceramarcae* and *M. unduaviensis*. The limits of range of these species are correlated with those of the East Andean montane forests between the upper reaches of the Rio San Franzisco and Rio Bermejo in the south and the Rio Marañon in the north. The lack of comparable forest biomes in the West Andean part of central and southern Peru (cf. KOEPCKE 1957), together with the isolating effect of the Eastern Cordillera, explains why

Fig. 39. Ranges of typical Yungas faunal elements.

33 bird families reach their western limit at the eastern margin of the Yungas centre (cf. NIETHAMMER 1953). On the other hand all the families present in the West Andean part of Peru can also be found in the East Andean part except for the Thinocoridae, Rheidae, Phoenicopteridae and Recurvirostridae which occur in the Puna centre.

BOND & MEYER DE SCHAUENSEE (1942–43), NIETHAMMER (1953–56) and FORSTER (1958) have given a partial description of the birds and Lepidoptera. FORSTER (1958) regarded the lepidopterous fauna as: 'eine der interessantesten und wohl die artenreichste Südamerikas', but gave no precise figures. BOND & MEYER DE SCHAUENSEE (1942–43) and NIETHAMMER (1953–56) tried to divide the montane forest biome into three altitudinal zones. These were the Montana or tropical zone from 0–1000 m. in height; the Medio Yungas or subtropical zone from 1000–2500 m.; and the cloud forest or Ceja de Montana from 2500–3500 m. Compare in this connection also ELLENBERG (1958, 1959), HERZOG (1923), WEBERBAUER (1911, 1945).

The Montana zone does not belong to the Yungas centre as here defined, since it is used as a breeding area only by faunal elements of the Amazon centre. The Yungas faunal elements which have so far been collected in this zone are all birds and were taken outside the true breeding season. Furthermore I consider that the transition from the Montana zone to the Medio Yungas does not lie at 1000 m. but at a higher altitude. This is suggested by numerous lowland species with strongly developed gonads that NIETHAMMER collected higher than 1000 m. It is not yet possible to say how clear the separation between the Medio Yungas and the cloud forest region really is. It is stated, for example, that *Papilio isidorus* (Lepidoptera, Papilionidae) is strictly adapted to the Medio Yungas while *Papilio warscewiczii* only occurs in the cloud forest. However the available locality data for the reptiles, amphibians, mammals and birds show no clear separation between the two zones (data in: COPE 1876, BOULENGER 1900, BARBOUR & NOBLE 1920, BURT & BURT 1933, PEARSON 1951, 1957, CABRERA 1957–1961). Seasonal migrations, like those seen in the Serra do Mar certainly occur here also (NIETHAMMER 1953–56). At present there is little information about their extent.

98 bird species and 24 mammals can be reckoned as faunal elements of the Yungas centre on the basis of data in ZIMMER (1931–1955), CABRERA (1957–1961), KOEPCKE & KOEPCKE (1963) and MEYER DE SCHAUENSEE (1966, 1970). The meagre information on reptiles and amphibia cannot be taken as representative.

204 species of breeding birds are at present known from the East Peruvian and Bolivian montane forests. This indicates, by comparison with the number of endemic species, that only 48% of the species that occur can be taken as faunal elements of the centre. These faunal elements have their closest connections with the montane forest biome of the Ecuadorian and Colombian Andes. Such a relationship can be illustrated by the range of the Tinamid genus *Nothocercus* (fig. 14). *N. nigrocapillus* can be taken as a faunal element of the Yungas centre and *N. julius* as a faunal element of the Colombian montane forest centre. The third species, *N. bonapartei*, is polytypic and exists as different endemic sub-

species in the Venezuelan, Colombian and Central American montane forest centres. A whole series of monotypic species occurs both in the Colombian montane forest centre and the Yungas centre and clearly shows the close relationship of the two biomes e.g. *Mazama bricenii.*

Of the 98 monocentric bird species at least 57 have their closest relatives in the Colombian montane forest centre. A direct connection with the Pantepui centre is indicated by *Aulacorhynchus derbianus* (cf. fig. 30). 17 monocentric species of birds have their closest relatives in the Amazon centre. SCHALLER (1958) remarks that 'the soil fauna in the cloud forest and the high plains of the Andes has a greater role in the biology of productivity'* than in the hot and humid lowland rain forest.

---

* 'Bodenfauna in den Nebelwäldern und auf den Hochebenen der Anden eine größere Produktions-biologische Rolle spielt.'

27. The Puna centre

The position of the centre (fig. 5, no. 27) is defined by the ranges of *Pterocnemia pennata garleppi*, *Lama vicugna*, *Tinamotis pentlandii*, *Felis jacobita*, *Hippocamelus antisiensis*, *Conepatus rex rex*, *Plegadis ridgwayi* and *Phoenicoparrus jamesi* (fig. 40). The centre is named after the high plain of the region. Expansive faunal elements of the Puna centre penetrate the Peruvian Andes subcentre of the North Andean centre. This penetration probably happened only recently, since the populations concerned are not morphologically differentiated. Eco-

Fig. 40. Types of distribution of Puna faunal elements. *Conepatus rex rex* is closely related to the subspecies *C. r. inca* of the Andean Pacific centre.

logically speaking the Peruvian Andes subcentre is a transition between the Puna centre and the Bogotá subcentre (cf. also CABRERA 1957).

Geographically speaking the region represents an enormous elevated mass which has had a modifying and improving effect on the climate (cf. MONHEIM 1956). It is surrounded on all sides by the mountain chains of the Andes and represents the broadest part of the Andean range. Its present-day altitude mainly results from elevation during the Pliocene (TROLL 1927, 1928, 1931, 1968, MACHATSCHEK 1955, DE LATTIN 1967).

Of all the birds that occur in South America, 35 species must be reckoned as faunal elements of the Puna centre. They are:

Pterocnemia pennata garleppi
Tinamotis pentlandii
Podiceps taczanowskii
Plegadis ridgwayi
Phoenicoparrus andinus
Phoenicoparrus jamesi
Anas puna
Buteo poecilochrous
Fulica gigantea
F. cornuta
Charadrius alticola
Phegornis mitchellii
Gallinago andina
Recurvirostra andina
Metriopelia aymara
Cinclodes atacamensis
Schizoeca helleri
Asthenes wyatti

A. maculicauda
Muscisaxicola rufivertex
M. juninensis
Petrochelidon andecola
Mimus dorsalis
Anthus furcatus
Sicalis lutea
S. uropygialis
Diuca speculifera
Idiopsar brachyurus
Compsospiza garleppi
Phrygilus atriceps
P. fruceti
P. dorsalis
P. erythronotus
Spinus crassirostris
S. atratus

It is striking that only six of these species (17.1%) belong to families regarded by MAYR (1964) as South American in origin. These are:

Pterocnemia pennata garleppi (Rheidae)
Tinamotis pentlandii (Tinamidae)
Cinclodes atacamensis (Furnariidae)

Schizoeca helleri (Furnariidae)
Asthenes wyatti (Furnariidae)
Asthenes maculicauda (Furnariidae)

The families to which the other species belong in all likelihood came in from the north—ten of the genera cited also occur in central Europe.

The origin of the 39 mammalian faunal elements in the Puna centre is not yet clear in all cases. Discussion of the problems involved can be found in GRANDI-DIER & NEVEU-LEMAIRE (1908), PEARSON (1948, 1951, 1957, 1960), KOFORD (1957), HERSHKOVITZ (1969) and SIMPSON (1969). Because of the uncertainties I shall not try to acribe the mammals to regions of origin. However it is interesting that 18 of the faunal elements belong to the Cricetidae which are widely distributed as a family in the Holarctic region. The rodents are represented by a total of 24 faunal elements in the Puna centre:

Thomasomys ladewi (Cricetidae)
T. oreas (Cricetidae)

93

Akodon andinus lutescens (Cricetidae)
A. pacificus (Cricetidae)
A. puer (Cricetidae)
A. amoenus (Cricetidae)
A. berlepschii (Cricetidae)
A. jelskii [with nine subspecies (Cricetidae)]
Oxymycterus paramensis (Cricetidae)
Calomys frida (Cricetidae)
C. lepidus [with seven subspecies (Cricetidae)]
Phyllotis caprinus (Cricetidae)
P. osilae [with five subspecies (Cricetidae)]
P. sublimis (Cricetidae)
Andinomys edax (Cricetidae)
Chinchillula sahamae (Cricetidae)
Punomys lemminus (Cricetidae)
Neotomys ebriosus (Cricetidae)
Ctenomys opimus (Ctenomyidae)
C. peruanus (Ctenomyidae)
Lagidium peruanum (Chinchillidae)
Chinchilla brevicauda (Chinchillidae)
Microcavia niata (Caviidae)
Galea musteloides musteloides (Caviidae)

There are 27 Puna faunal elements among the amphibia. Of these only *Bufo spinolosus spinolosus* and the South American subspecies related to it have come in from the north. Unlike *Bufo marinus*, *B. spinolosus* must be considered as a new immigrant cf. MÜLLER (1968). The same holds for *B. veraguensis* O. SCHMIDT (1857, cf. SAVAGE 1969, GALLARDO 1961) although this is an exceptional species among New World Bufonids, with many morphological peculiarities such as the lack of a tympanum.

The high percentage of endemic species of *Telmatobius* in the centre is worth mentioning. According to GORHAM (1966) and CEI (1969) they represent 50% of all known species of the genus, and can be listed as follows:

Telmatobius albiventris   *T. jelskii*
T. brevirostris   *T. latirostris*
T. crawfordi   *T. marmoratus*
T. culeus   *T. oxycephalus*
T. hauthali   *T. rimac*
T. ignavus   *T. simonsi*
T. intermedius

Of these 13 species *T. albiventris*, *T. culeus* and *T. marmoratus* are endemic to Lake Titicaca. These three species of water frog are specially noteworthy in connection with the problem of intralacustrine speciation in the lake (cf. BROOKS 1950, KOSSWIG & VILLWOCK 1964). Of all known *Telmatobius* species they have the greatest tendency to subspeciation. Thus there are eight known subspecies of *T. marmoratus*, six of *T. culeus* and four of *T. albiventris*.

Among the reptiles the 12 faunal elements of the Puna centre at present known are all of South American origin. The number of species of the Iguanid *Liolaemus* recognised from the centre has decreased by five following the work of

94

HELLMICH (1962). He pointed out that *L. lenzi* of BOETTGER (1891), *L. annectens* and *L. tropidonotus* of BOULENGER (1901, 1902) and *L. variabilis* and *L. bolivianus* of PELLEGRIN (1909) were all synonyms of *L. multiformis*. The latter is the commonest lizard of the Puna centre according to PEARSON (1954).

Palaeontological results show that a lowland rain-forest flora was able to develop in the region of the Puna centre in the Miocene, but disappeared at the end of the Pliocene (MENENDEZ 1969).

Some Puna faunal elements may derive from pre-adapted lowland ancestors which migrated into the new upland environment. Others may be descended from taxa that were lifted up into a cooler climation-zone by the Andean orogenesis. Unfortunately studies of present-day distribution do not always clearly distinguish the two possible cases and only a continuous fossil history will be decisive. The occasional palaeontological results do suggest, however, that both modes of origin are indeed possible. In addition newer immigrants can be recognised as such by studying their present-day ranges.

The closest relationships of the faunal elements of the Puna centre are with the North Andean and the Patagonian centres (cf. fig. 41). Thus of the 35 species of birds that can be reckoned as Puna faunal elements, 14 have their closest related populations in the Patagonian centre and 15 in the North Andean centre. In this connection compare in fig. 41 the ranges of *Pterocnemia pennata garleppi* (Puna faunal element) and *P. p. pennata* (Patagonian faunal element) and also *Tinamotis pentlandii* (Puna faunal element) with *T. ingoufi* (Patagonian faunal element).

An exceptional case is *Plegadis ridgwayi* which occurs in both the Puna and the Andean Pacific centres. The two populations are not subspecifically distinct however and gene exchange obviously still exists between them. The range of this species indicates a type of connection which is commoner among mammals, amphibians and reptiles than among birds (cf. the ranges of *Bufo spinolosus* and *Conepatus rex* in fig. 40). A connection with the Andean Pacific centre is indicated by six of the mammals of the centre, four of the amphibians, and six out of the twelve reptiles.

Birds that breed in the Alaskotundral, Neotundral and Atlantotundral centres of the Nearctic region (DE LATTIN 1967) prefer to pass the northern winter as migrants in the Puna and Patagonian centres (JOHANSEN 1969).

If the regional concept were applied consistently my results show that the boundary of the Nearctic region above the tree-line, at least for the birds, would lie not in North America, nor in Central America, but in South America. This also holds for the mammals as shown for example by the high proportion of Cricetids in the Puna centre. It is not true for the poikilothermous amphibians and reptiles.

The Lepidoptera of the Puna centre show much the same biogeographical relationships as the birds. FORSTER (1958, p. 845) remarks that the lepidopterous fauna of the unforested areas above the tree-line has: 'very few South American elements and mainly consists of genera which are principally Holarctic in distribution like the genus *Colias*, or at least have their main relatives in the

Fig. 41. Distributions in the Puna and Patagonian centres. The Puna faunal elements *Pterocnemia pennata garleppi*, *Lama vicugna* and *Tinamotis pentlandii* are replaced by closely related species or subspecies in the Patagonian centre. *Lama guanaco* and *Phoenicopterus chilensis* are polycentric species which occur both in the Puna and in the Patagonian centre.

Holarctic region like the Pierid genus *Phulia*'.* His results have been confirmed and supplemented by the work of BREYER (1936, 1939), URETA (1936–1937), TURK (1955) and HOVANITZ (1945, 1958). TURK (1955) has shown that there are for example endemic species of chilopods that inhabit both the Puna and North Andean centres. KRAUS (1954) remarks that for the poorly studied diplopods zoogeographical conclusions cannot yet be drawn.

* 'sehr wenig südamerikanische Elemente enthält, dagegen in der Hauptsache Gattungen, die vorzugsweise holarktisch verbreitet sind, wie z.B. die Gattung *Colias*, oder die wenigstens ihre nächsten Verwandten in der Holarktis haben, wie die Weisslingsgattung *Phulia*.'

28. The Marañon centre

The position of this inter-Andean centre (fig. 5, no. 28) can be defined by the ranges of *Melanopareia maranonica* (cf. fig. 42), *Pseudogonatodes barbouri* and *Ameiva bifrontata concolor*. The closest faunistic connections are with the Andean Pacific centre. These connections are shown by the following list of the bird fauna; in the list (1) after the subspecies name indicates that the subspecies occurs in the Ecuadorian subcentre of the North Andean centre; (2) indicates occurrence in the Peruvian subcentre.

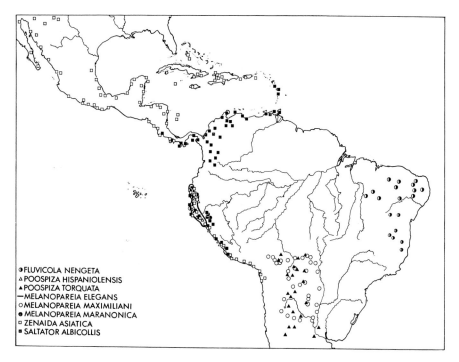

FLUVICOLA NENGETA
POOSPIZA HISPANIOLENSIS
POOSPIZA TORQUATA
MELANOPAREIA ELEGANS
MELANOPAREIA MAXIMILIANI
MELANOPAREIA MARANONICA
ZENAIDA ASIATICA
SALTATOR ALBICOLLIS

Fig. 42. Connections of Andean Pacific and Marañon faunal elements. *Saltator albicollis* occurs as the same subspecies in the Andean Pacific and in the Marañon centres. It is represented by other subspecies in the Magdalena, Cauca, Barranquilla and Caribbean centres. *Zenaida asiatica* is distributed on the one hand in Andean Pacific centre, and on the other as particular subspecies in the Central American Pacific centre, in the Yucatan centre, in the Coco centre and on the West Indian islands. *Melanopareia maranonica* is a Marañon faunal element; its closest relative is *M. elegans* which is an Andean Pacific faunal element. Both these species can be derived from *M. maximiliani* which is a Chaco faunal element. *Poospiza hispaniolensis* is the closest relative of *Poospiza torquata*. *Fluvicola* occurs as distinct subspecies in the Caatinga and in the Andean Pacific centre.

Andean Pacific centre	Marañon centre
Aratinga rubrolarvata	*A. r.*
A. frontata	*A. f.*
Psittacula coelestis	*P. c.*
Brotogeris pyrrhopterus	*B. p.*
Veniliornis callonotus major (1)	*V. c. major*
V. callonotus callonotus (2)	—
Thamnophilus bernardi bernardi (1)	*T. b. bernardi*
T. b. piurae (2)	—
Melanopareia elegans elegans (1) ⎫	*M. maranonicus*
Melanopareia elegans pauculensis (2) ⎬	
Furnarius cinnamomeus	*F. c.*
Synallaxis stictothorax stictothorax (1)	*S. s. chinchipensis*
Synallaxis stictothorax maculata (2)	—
Phaeomyias murina tumbezana (1)	*P. m. tumbezana*
P. murina inflava (2)	—
Hapalocercus melorhyphus fulviceps	*H. m. fulviceps*
Camptostoma obsoletum sclateri	*C. o. sclateri*
Myiophobus fasciatus crypterythrus	*M. f. crypterythrus*
Pyrocephalus rubinus heterurus	*P. r. saturatus*
Myiochanes brachytarsus punensis	*M. b. punensis*
Heleodytes fasciatus pallescens (1)	*H. f. pallescens*
H. fasciatus fasciatus (2)	—
Mimus longicaudus punensis (1)	—
M. longicaudus longicaudus (2)	*M. l. longicaudus*
Turdus reevei	*T. r.*
Sicalis flaveola valida	*S. f. valida*
Arremon abeillei	*A. nigriceps*

This comparative table shows clearly that the Marañon centre has closer affinities with the Ecuadorian subcentre than with the Peruvian subcentre. The reptiles of the centre show the same thing. Thus the opisthoglyph *Leptodeira septentrionalis* occurs as the subspecies *L. s. larcorum* both in the Marañon centre and the Ecuadorian subcentre.

Ten species shown in the list have their closest relatives east of the Andes. This is true for *Melanopareia, Furnarius, Phoeomyias, Hapalocercus, Camptostoma, Myiophobus, Pyrocephalus, Sicalis, Myiochanes* and *Mimus*.

In the Urubamba valley, which lies between the Marañon valley and the nearest Bolivian campo, 66 species of birds occur (CHAPMAN 1921). Besides 38 very widespread species, which live preferentially or exclusively in unforested biotopes, 19 species are of definite Brazilian origin (Campo Cerrado or Caatinga centre). It is not likely that these forms immigrated accidentally.

Two Marañon populations i.e. *Saltator striaticollis flavidicollis* and *Pheugopedius paucimaculatus*, have closer connections with the arid regions of Colombia than with those of the Pacific coast of Peru.

Besides the endemic species *Melanopareia maranonica*, two further endemic species inhabit the Marañon centre. These are: *Columba oenops* and *Incaspiza watkinsi*. *C. oenops* is related to *C. corensis* of the Caribbean centre. However the species closest related to *I. watkinsi* is *I. laeta* which also occurs in the Marañon valley, though indeed above 3300 m.

98

None of the other inter-Andean valleys has the importance of the Cauca, Magdalena or Marañon centres, and they have had no special role except as staging posts in the movement of open-landscape species. This is just as true for the Dabeiba and Dagua valleys on the west side of the Western Cordillera as for the Cucuta valley, south of the Catatumbo centre. Savanna forms occur in these valleys but have not evolved endemic species in them.

The only exception is the Chicamocha valley from which only the bird fauna is well known at present (BORRERO & OLIVARES 1955, BORRERO & HERNANDEZ 1958). From the studies available it appears that this valley has a number of endemics i.e. *Amazilia castaneiventris, Thryothorus nicefori, Thamnophilus multistriatus oecotonophilus, Catharus aurantiirostris inornatus, Arremon schlegeli canidorsum*. These forms are closely related to populations in the Magdalena centre.

29. The Andean Pacific centre

The centre (fig. 5, no. 29) can be defined by the ranges of the monotypic Trochilid genus *Rhodopis* and of the ground iguana *Tropidurus peruvianus* (fig. 43). Three subcentres can easily be separated from each other. The first of these

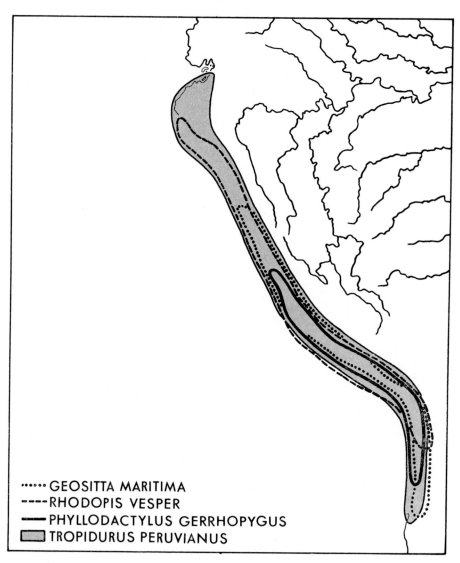

Fig. 43. Ranges of Andean Pacific faunal elements.

is the **Ecuadorian subcentre,** which can be defined by the ranges of: *Piezorhina cinerea, Gnathospiza taczananowskii, Cyanocorax mystacalis* (fig. 44), *Spinus siemiradzkii,* the Teiid *Dicrodon guttulatum, Bothrops barnetti* and *Micrurus mertensi* (fig. 45). The second subcentre is the **Peruvian subcentre.** It has a more

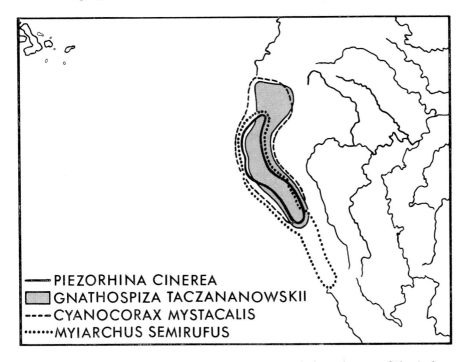

Fig. 44. Distribution of four faunal elements of the Ecuadorian subcentre of the Andean Pacific centre.

arid climate than the Ecuadorian subcentre and can be defined by the ranges of: *Paralomys gerbillus* (monotypic Cricetid genus), *Bothrops roedingeri, Ctenoblepharis adspersus* (Iguanidae), *Thaumastura cora* (monotypic genus), *Atlapetes nationi, Xenospingus concolor* (monotypic genus) and the Teiid *Dicrodon heterolepis* (fig. 45). The third and southernmost subcentre is the **Chilean subcentre** which can be defined on the ranges of: *Ctenomys robustus* (Ctenomyidae), *Octodon degus* (Octodontidae), *Mimus thenca, Callopistes maculatus* and *Pteroptochos megapodius* (fig. 46, cf. also fig. 83).

In dealing with the Marañon centre I have already remarked on the close relationship between it and the Andean Pacific centre. An exact analysis shows however that, with few exceptions, the Marañon centre has its closest affinities with the Ecuadorian subcentre. These close affinities, which are obvious from a tabulation of the birds of the Marañon centre, show themselves by the fact that some subspecifically distinct populations of the Marañon centre very largely agree with those of the Ecuadorian subcentre but differ from those of the Peruvian subcentre. *Mimus longicaudatus* is an exception. The nominate subspecies occurs in the Peruvian subcentre and also in the Marañon centre.

Some Andean Pacific faunal elements are differentiated into subspecies. They

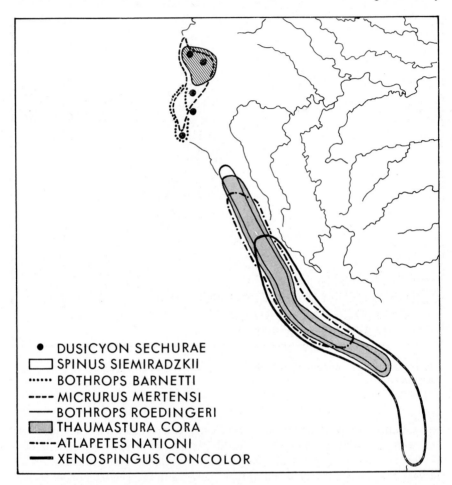

● DUSICYON SECHURAE
▭ SPINUS SIEMIRADZKII
······ BOTHROPS BARNETTI
---- MICRURUS MERTENSI
—— BOTHROPS ROEDINGERI
▦ THAUMASTURA CORA
-·-·-ATLAPETES NATIONI
━━ XENOSPINGUS CONCOLOR

Fig. 45. Ranges of faunal elements of the two northernmost subcentres of the Andean Pacific centre i.e. Ecuadorian subcentre: *Dusicyon sechurae, Spinus semiradzkii, Bothrops barnetti* and *Micrurus mertensi*; and Peruvian subcentre: *Bothrops roedingeri, Thaumastura cora, Atlapetes nationi* and *Xenospingus concolor.*

show that the subcentres obtained by analysing the smallest geographical ranges have acted as centres of subspecific differentiation. Thus in the Ecuadorian subcentre *Heleodytes fasciatus* occurs as *H. f. pallescens* but in the adjacent Peruvian subcentre it is represented by *H. f. fasciatus.*

—OCTODON DEGUS
● MIMUS THENCA
·····PTEROPTOCHOS ME-
GAPODIUS

Fig. 46. Distribution of three faunal elements of the Chilean subcentre of the Andean Pacific centre (*Octodon degus, Mimus thenca, Pteroptochos megapodius*).

The following bird species are to be reckoned as faunal elements of the Andean Pacific centre following MEYER DE SCHAUENSEE (1966, 1970) and KOEPCKE (1970). Monotypic genera are indicated by (m):

Burhinus superciliaris
Columbina cruziana
Leptotila ochraceiventris
Aratinga erythrogenys
Forpus coelestis
Brotogeris pyrrhopterus
Leucippus baeri
Rhodopis vesper (m)
Thaumastura cora (m)
Geositta maritima
Cinclodes nigrofumosus
Myiarchus semirufus
Myiopagis subplacens
Phaeomyias leucospodia

Cyanocorax mystacalis
Thryothorus superciliaris
Dives warszewiczi
Basileuterus fraseri
Piezorhina cinerea (m)
Sporophila peruviana
Gnathospiza taczanowskii
Atlapetes nationi
A. albiceps
Rhynchospiza stolzmanni (m)
Xenospingus concolor (m)
Poospiza hispaniolensis
Spinus siemiradzkii
Sporophila telasco

The lack of Tinamids among the faunal elements of the Andean Pacific centre is very striking. The ranges of the various *Geositta* species in the centre show

103

that this genus of birds is a sensitive indicator for different biotopes (KOEPCKE 1967). Thus *Geositta crassirostris* occurs as the nominate form in central Peru round Lima and Ayacucho in the coastal Lomas (50–800 m.). In the hill savanna of the western slopes of the Andes (1500–3000 m.) on the other hand the sub-species *G. c. fortis* is found. The two subspecies are separated from each other by a strip of extreme desert.

The ground iguana *Tropidurus* is very richly represented in the Andean Pacific centre in contrast with the unforested biomes of Brazil. In some cases it is as good a biotope indicator—of sandy coasts, rocky coasts etc.—as the species of *Geositta* (MERTENS 1956). The distribution of the six known species and four subspecies of *Tropidurus* corresponds to the division of the centre into the sub-centres already discussed i.e. *T. peruvianus*, *T. occipitalis*, *T. tarapacensis*, *T. theresiae*, *T. theresioides*, *T. thoracicus*. As with *Geositta*, some *Tropidurus* species, including *T. peruvianus* are divided into subspecies whose ranges correspond to zones of altitude.

The occurrence of the Boid genus *Trachyboa* in the Ecuadorian subcentre is worthy of note. This genus has only two species of which one is a faunal element of the Choco subcentre of the Colombian Pacific centre (*T. boulengeri*), while the other is an Ecuadorian faunal element (*T. gularis*).

The affinities of the Andean Pacific centre to the other Neotropical centres are shown in fig. 42. A very close connection between the centre and the central Brazilian centres is indicated by a number of forms i.e.: *Cyanocorax mystacalis*, *Synallaxis tithys*, *Sakesphorus bernardi*, *Melanopareia elegans*, *Columbigallina talpacoti buckleyi*, *Furnarius leucopus cinnamomeus*, *Euscarthmus melorhyphus fulviceps*, *Fluvicola nengeta atripennis* and *Sicalis flaveola valida*. In this way the relationship between the Andean Pacific centre, on the one hand, and the Chaco and Caatinga centres, on the other, is made particularly obvious.

Thus the range of *Fluvicola nengeta* shows a gross disjunction. Populations of this species inhabit the Ecuadorian subcentre and also the arid regions of the Brazilian Caatinga centre. A relationship to the Chaco centre is shown by the ranges of *Poospiza* and *Melanopareia*. Thus *P. hispaniolensis* occurs in the Andean Pacific centre while *P. torquata*, which is the species closest related to it, occurs in the Chaco centre. *Melanopareia elegans* is an Ecuadorian element while the closest related species *M. maranonica* is a Marañon faunal element; but the species closest related to those two is *M. maximiliani* of the Chaco centre.

There are also close connections with the arid areas of the Caribbean centre and, in Central America, to the Barranquilla, Central American Pacific and Yucatan centres. Compare in this connection the ranges of *Saltator albicollis* and *Zenaida asiatica* in fig. 42.

The gecko genus *Phyllodactylus* has a similar range to *Zenaida asiatica* in South and Central America. Twelve species are faunal elements of the Central American Pacific centre: one is a Yucatan and one a Caribbean faunal element, and thirteen species are Andean Pacific faunal elements. Six other species live in the Antilles and on the Galapagos Archipelago, as discussed under the Galapagos centre. These are: *Phyllodactylus gerrhopygus, Ph. heterurus, Ph. inaequalis, Ph. lepidopygus, Ph. microphyllus* and *Ph. reissii.*

The fact that the Andean Pacific centre is so rich in endemics (see also OLROG, 1969), which is true for the cacti as it is for the vertebrates (BUXBAUM, 1969), shows that the centre has existed for a long time and been well isolated. It is said that the region was already arid in character during the Pliocene (MENENDEZ 1969).

CHAPMAN (1926) supposed that *Zenaida asiatica meloda* and *Dives warscewiczi* reached the Andean Pacific centre from the north at a time when the forest areas of the Choco subcentre did not yet exist—purportedly in the Pliocene. This supposition is highly unlikely, however, since the level of differentiation of populations that are now disjunct does not suggest such a great age. It is much more likely that these forms came in during a warm period of the Pleistocene or in post-glacial times by way of the dry valleys of the interior of the Andes, where they still occur.

On the other hand the monotypic genera may have been indigenous to this area in pre-glacial times, at least in part. Concerning this however, nothing definite can be said without fossil evidence. Conceivably, indeed, the great abundance of endemics signifies that part of the fauna originated before the rise of the Andes (cf. the **Nothofagus** centre). However, in these cases also, nothing further can be said without finding fossils, since forms are known which have successfully crossed the Andes in post-glacial and modern times (cf. also the Galapagos fauna).

30. The Galapagos centre

The inventory of the plants and animals of this group of islands (fig. 5, no. 30), which are only 7840 km². in extent, can be taken as essentially complete: (HOOKER 1847, STEWART 1911, DENBURGH & SLEVIN 1913, DALL & OSCHNER 1928, FOWLER 1944, SWARTH 1929, CHUBB 1933, KINSEY 1942, SVENSON 1946, LACK 1947, BOTT 1958, BOWMAN 1961, 1966, EIBL-EIBESFELDT 1962, 1966, BROSSET 1963,

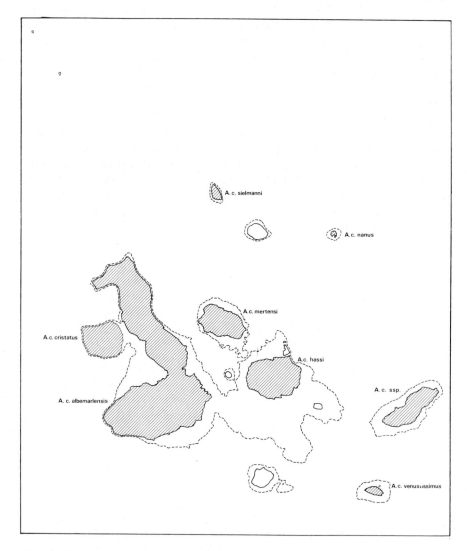

Fig. 47. The range of the *Iguanid Amblyrhynchus cristatus* on the Galapagos. The dashed line is the 200 m. isobath. The islands on which *A. cristatus* occurs are cross-hatched.

CARPENTER 1964, MERTENS 1960, 1963, J. NIETHAMMER 1964, BAILEY 1968, CURIO 1969, LELEUP & LELEUP 1968, 1970).

The vertebrate fauna entirely lacks fresh-water fishes and amphibians. The animals that do exist have very close connections with the Andean Pacific and Central American Pacific centres and weaker connections with the Antilles. I shall illustrate these relationships among the reptiles first. *Amblyrhynchus cristatus*, which belongs to a monotypic genus, occurs as closely related sub-species on the islands of Narborough, Albemarle and Indefatigable, which all lie within the 200 m. isobath (fig. 52 for names of islands). On James is found the more strongly distinct subspecies *A. c. mertensi* (fig. 47). The populations most strongly differentiated from the nominate form are *A. c. nanus* on Tower and *A. c. venustissimus* on Hood. The varying degrees of differentiation of the populations outside the 200 m. line and their distributions, can be explained by passive drifting, and the same is true for the genus *Conolophus*. In qualification it must be pointed out that the animals use their swimming ability only to a limited extent and are water-shy (EIBL-EIBESFELDT 1962), perhaps because of the numerous sharks round the coast. However BARTHOLOMEW (1966) and HOBSON (1969) consider that *Amblyrhynchus cristatus* only enters the water for short periods because its preferred temperature of 35° to 37°C is about 10°C higher than the water temperature. In the waters of the Galapagos Hobson observed *Triaenodon obesus* and *Carcharinus* sp. which were paying no attention to the *Amblyrhynchus*.

The strong differentiation of *Conolophus* and *Amblyrhynchus*, both monotypic genera, indicates that they immigrated very early, perhaps in the Pliocene or Pleistocene. The closest relatives of both genera occur in Central America and the Antilles (*Ctenosaurus, Iguana, Cyclura*). According to AVERY and TANNER (1971), *Conolophus* and *Amblyrhynchus* are closest to *Iguana* (fig. 48).

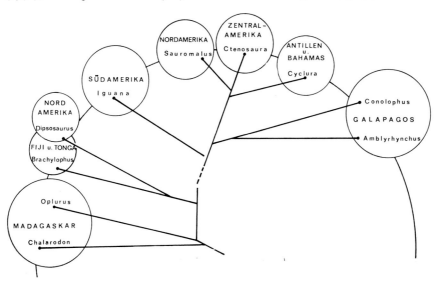

Fig. 48. Evolutionary scheme of the Iguanidae (after AVERY & TANNER 1971).

The remainder of the Galapagos reptiles cannot be separated generically from the populations on the mainland. Their closest relatives are faunal elements of the Andean Pacific region.

Like *Amblyrhynchus* the species of *Tropidurus* clearly indicate the close affinity of the central group of islands of Narborough, Albemarle, James and Indefatigable. This is obvious from their distribution, their degree of morphological differentiation and their pattern of behaviour (CARPENTER 1964, EIBEL-EIBESFELDT

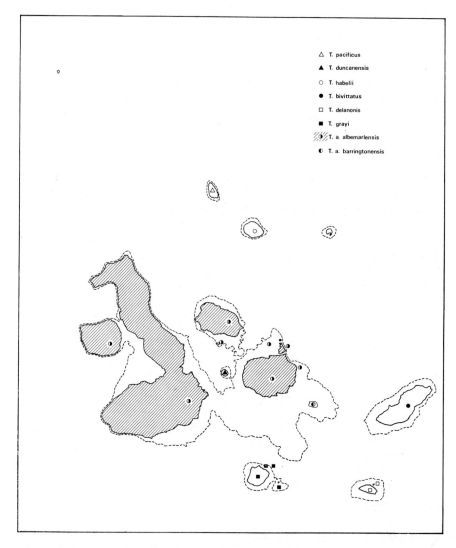

Fig. 49 Distribution of the Iguanid genus *Tropidurus* on the Galapagos. The range of *T. albemarlensis* is cross-hatched to bring out the close relationship of the central group of islands within the 200 m. line.

1966); see, in this connection, the distribution of the nominate subspecies of *T. albemarlensis* in fig. 49. The island of Barrington, which is likewise within the 200 m. line, still shows close affinities with the central group of islands, although some distance from Indefatigable. *Tropidurus albemarlensis* occurs on it as a distinct subspecies. By contrast all the islands outside the 200 m. isobath have populations which are morphologically clearly distinct and which, despite their allopatric distribution, are given specific rank. The existence of *Tropidurus* on the various islands (fig. 49) in no way indicates a one-time land connection,

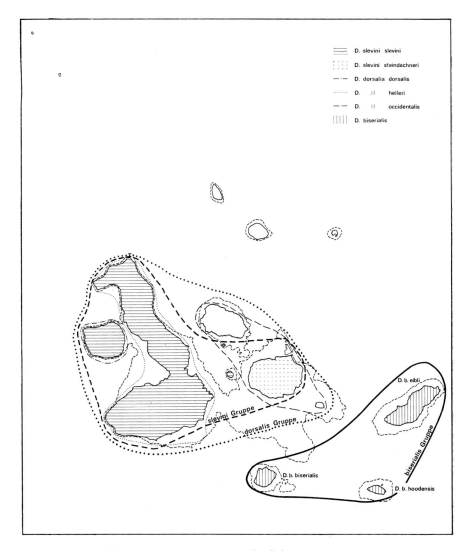

Fig. 50. Range of the snake genus *Dromicus* on the Galapagos.

contrary to what DENBURGH & SLEVIN (1913) etc. supposed. It can readily be explained by passive drifting (cf. discussion of Caribbean centre). The most closely related species of *Tropidurus* occur in the Andean Pacific centre.

The close relationship of the central islands with each other is also apparent from the snake genus *Dromicus* (fig. 50). The central islands are inhabited by the *dorsalis* group, as three subspecies, and by the *slevini* group as two subspecies (MERTENS 1960), and, except for James, Barrington and Brattle the distribution is sympatric. On Charles, Hood and Chatham the *biserialis* group occurs.

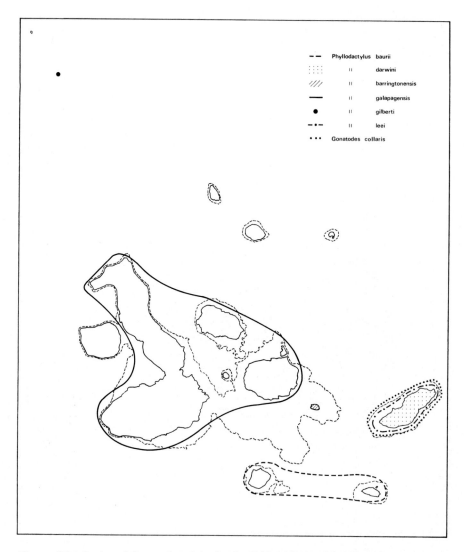

Fig. 51. Distribution of the species of the family Gekkonidae on the Galapagos.

Distribution of *Dromicus* species on the Galapagos, after MERTENS (1960):

Dromicus biserialis biserialis	Charles and Gardner near Charles.
D. biserialis hoodensis	Hood and Garner near Hood.
D. biserialis eibli	Chatham.
D. dorsalis dorsalis	James, Jervis, Barrington, Indefatigable, South Seymour.
D. dorsalis occidentalis	Narborough.
D. dorsalis helleri	Brattle and Albemarle.
D. slevini slevini	Duncan, Albemarle, Narborough.
D. slevini steindachneri	Jervis, South Seymour, Indefatigable.

The Galapagos snakes are closest related to the Andean Pacific faunal element *Dromicus chamissonis*, which has scale pits like those characteristic of the Galapagos forms.

The geckos of the Galapagos have the same pattern of distribution as the other reptiles (fig. 51). The distribution of *Phyllodactylus galapagensis* shows the close affinity of the central islands. The lack of gecko species on Narborough is very striking.

There is a decline in the number of species of geckos from the east, on Chatham, to the west on Wenman. Thus on Chatham the three species *Phyllodactylus darwini*, *P. leei* and *Gonatodes collaris* occur, whereas on all the other islands there is never more than one single species of the genus *Phyllodactylus*. This does not so much reflect the stronger ecological differentiation of Chatham as indicate the direction of immigration of the Galapagos geckos, whose closest relatives occur in the Andean Pacific centre (*Phyllodactylus*). However, while *Phyllodactylus* is widely distributed in the Old World (UNDERWOOD 1954, WERMUTH 1965), *Gonatodes* belongs to the purely Neotropical subfamily Sphaerodactylinae. The 17 species of *Gonatodes* (WERMUTH 1965, VANZOLINI 1968, PETERS & DONOSO-BARROS 1970) have their centre of diversity in Central America and the northern part of South America, the genus being absent in the Serra do Mar centre. In addition to forms adapted to rain forest, such as *Gonatodes concinnatus*, there are also eurytopic species like *G. albogularis*. *Gonatodes caudiscutatus*, which CURIO rediscovered on Chatham (MERTENS 1963), was taken 'under stones in the highlands which are damp in climate during the Garua season'.*

The giant tortoises of the Galapagos, belonging to the *Geochelene elephantopus* group, can be derived from species on the mainland of South America. They share their generic name, formerly *Testudo*, with the South American forest species *G. denticulata*, and with *G. carbonaria* and *G. chilensis* in the open landscapes of South America (MÜLLER 1970), and these three forms are their closest relatives. *G. elephantopus* is represented by a different subspecies on almost every island in the Galapagos (fig. 52). This indicates two things. Firstly, the original population must have had great genetic plasticity. And secondly, the possibilities of drifting from island to island have been very restricted. The

* 'unter Steinen im zur Garua-Zeit feuchten Hochland.'

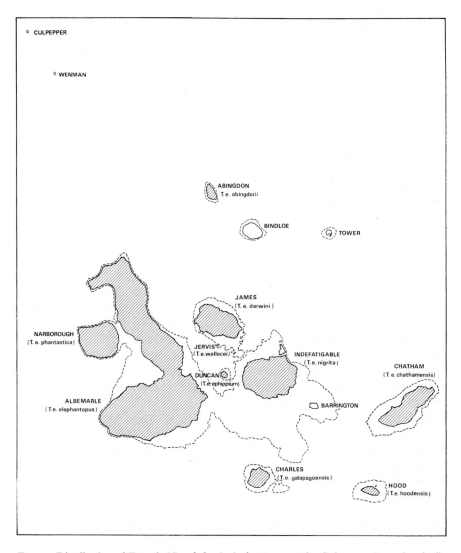

Fig. 52. Distribution of *Testudo* (*Geochelone*) *elephantopus* on the Galapagos (cross-hatched). Each island is inhabited by an endemic subspecies.

difficulties of such drifting are also demonstrated by the lack of buoyancy of the tortoises. This caused DENBURGH (1914), BEEBE (1924) and VINTON (1951) to suppose that the giant tortoises had come to the Galapagos along a Miocene continental land bridge from Central America by way of the Cocos islands. Considering the obviously rapid rate of subspecification of these tortoises, however, it would be remarkable, if that were true, that they have not differentiated further from their group of origin. It is more plausible to suppose that the ancestors of *Geochelone elephantopus* were more buoyant and reached the

Galapagos by passive drifting from South America. In point of fact *G. elephan-topus* moves well over water but uses up a great deal of energy by rapid paddling movements of its limbs. On the island of Marajó, near the mouth of the Amazon, the species *G. carbonaria*, which reaches a length of up to 60 cm., prefers to live in flood savanna (MÜLLER 1970) and the juveniles can move in water. In this connection the existence of the giant tortoise *Testudo cubensis* in the 'Pleistocene' of Cuba is worth mentioning. According to WILLIAMS (1950) it is of a 'size comparable to that of the Galapagos specimens' and it is closely related to *G. elephantopus*, *G. denticulata*, *G. carbonaria* and *G. chilensis*. In the Pliocene of Argentina (Monte Hermoso) the species *Testudo praestans* occurs. It has many characters of the recent species as well as those of *T. cubensis* and can be considered as near to the hypothetical ancestral form. The different degrees of differentiation of the genus *Testudo* in South and North America support the view that *G. elephantopus* reached the Galapagos at the end of the Pliocene or the beginning of the Pleistocene.

The 89 breeding birds of the Galapagos, including the sea birds, can all be regarded as passive immigrants. This is true for the flightless cormorant *Nannop-terum harrisi* and the penguin *Spheniscus mendiculus* as well as for the endemic sea gulls *Creagrus furcatus* and *Larus fuliginosus*, the mocking-thrushes of the genus *Nesomimus* and the Darwin finches, of which 14 species inhabit the Galapagos and Cocos islands. These finches, or buntings, have been the object of numerous studies (BOWMAN 1961, HAMILTON & RUBINOFF 1963, 1964, MACARTHUR & WILSON 1963, CURIO & KRAMER 1964, CURIO 1969, etc.) but I wish to discuss here only the question of what their nearest continental relatives are. They can, in fact, most easily be linked with the aberrant Emberizine genus *Tiaris* which occurs in open landscapes in Central and South America. If the degree of difference between the Darwin finches and the other bunting-like birds of South America were taken as a measure for subdividing the Emberizi-nae, then a whole series of further subfamilies would need to be set up in conti-nental South America. That is to say that the rank of an independent subfamily —the Geospizinae—for the Darwin finches is not justified.

The mammalian fauna of the Galapagos consists only of small mammals, except for the fur seal *Arctocephalus galapagoensis*. The following forms occur:

Lasiurus brachyotis	on most islands
L. cinereus	on most islands
Rattus rattus	on most islands
Mus musculus	on most islands
Megalomys curioi	Indefatigable
Nesoryzomys indefessus	Indefatigable
N. darwini	Indefatigable
N. swarthi	James
N. narboroughi	Narborough
Oryzomys galapagoensis	Chatham
O. bauri	Barrington

Lasiurus brachyotis can be regarded as a subspecies of the continental *L. borealis* while *Lasiurus cinereus* occurs as an identical form on the continent; there must be gene exchange between the mainland and Galapagos populations. *Rattus rattus* and *Mus musculus* were introduced in historical times by man.

The closest relatives of *Nesoryzomys* and *Oryzomys* can be found in the open landscapes of Central and South America. The presence of *Megalomys* on the Galapagos is of interest. Its occurrence was established by NIETHAMMER only in 1964. The species was found on Indefatigable in pellets produced by *Tyto alba punctatissima* and *Asio flammeus galapagoensis*; it may in fact already be extinct. The genus *Megalomys* was previously known only on the islands of Martinique, Barbados and Santa Lucia in the Antilles and may already be extinct in those places as well. The range is therefore a relict range, indicating a wider distribution in former times. The fact that the Galapagos form differs very little from those of the Antilles indicates that the Galapagos populations probably reached those islands during the Pleistocene.

The situation can be summed-up as follows: Results from systematic zoo-geography and evolutionary genetics indicate that no land contact with the continent is needed to explain the vertebrate fauna of the Galapagos; and this holds also for the invertebrate fauna (LINELL 1899, WILLIAMS 1911, 1926, WHEELER 1919, MORGAN 1920, BOTT 1958, DE LATTIN 1967). Based on the degree of differentiation two groups of immigrants can be distinguished which obviously differ in age:

1. There is a group which differs greatly in morphology from the present-day mainland population—*Amblyrhynchus* and *Conolophus*. This group probably already reached the Galapagos at the end of the Pliocene (perhaps coming from the north). The sea currents of the area in question must then have been different to what they are today. AVERY & TANNER (1971) state that '*Amblyrhynchus* and *Conolophus*, which are endemic to the Galapagos Islands, have been separated from the mainland genera for a long time, as indicated by their high degree of differences' (fig. 48).

2. There is a group of reptiles which, compared with the first group, differ relatively little from the populations of the mainland. These are of the genera *Tropidurus, Dromicus, Phyllodactylus* and *Gonatodes*. These probably first reached the Galapagos from the coast of Peru during the Pleistocene.

31. The Caatinga centre

The position of this centre (fig. 5, no. 31) is defined by the ranges of *Kerodon rupestris* (cf. CABRERA 1957–1961), *Crotalus durissus cascavella* (this subspecies of this South American rattlesnake is well adapted to extremely dry areas cf. fig. 86), *Bothrops erijthromelas, Xiphocolaptes falcirostris, Phaethornis gounellei, Spinus yarellii, Epicrates cenchria xerophilus, Leimadophis poecilogyrus xerophilus, Tropidurus semitaeniatus* and *Gymnodactylus geckoides geckoides* (fig. 53). The endemicity of the Caatinga centre is greater than that of the Chaco centre and in reptiles is 10% of the total. In the Chaco centre endemic amphibians are almost entirely lacking, but this does not hold to the same extent for the Caatinga centre. Thus even within the genus *Physalaemus*

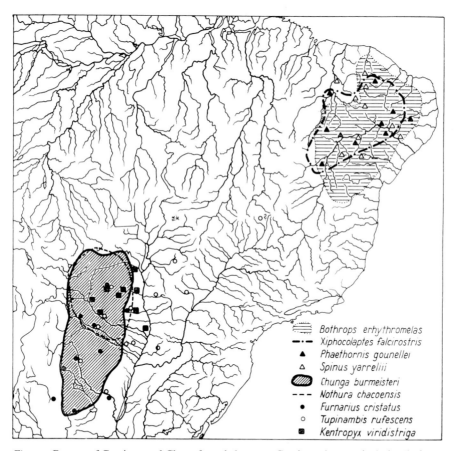

Legend:
Bothrops erhythromelas
Xiphocolaptes falcirostris
Phaethornis gounellei
Spinus yarellii
Chunga burmeisteri
Nothura chacoensis
Furnarius cristatus
Tupinambis rufescens
Kentropyx viridistriga

Fig. 53. Ranges of Caatinga and Chaco faunal elements. Caatinga elements include: *Bothrops erythromelas, Xiphocolaptes falcirostris, Phaethornis gounellei, Spinus yarellii.* Chaco faunal elements include: *Chunga burmeisteri, Nothura chacoensis, Furnarius cristatus, Tupinambis rufescens, Kentropyx viridistriga.*

there are three species that can be reckoned as Caatinga faunal elements i.e. *P. cicada*, *P. albifrons* and *P. kroyeri*, and there are two such elements in the Leptodactylidae i.e. *Odontophrynus carvalhoi* and *Pleurodema diplolistris*.

The closest relatives of the Caatinga faunal elements occur in the Chaco centre and the Campo Cerrado centre. The connection between the Chaco and the Caatinga centres becomes particularly obvious when a species range that includes both centres can be divided into subspecies. In such cases the ranges of the subspecies correlate with the two centres cf. the ranges of *Crypturellus tataupa*, (fig. 54) *Crotalus durissus* and *Phyllopezus pollicaris* (fig. 54) and *Columba picazuro* in fig. 55. The range of *Fluvicola nengeta* shows that the Caatinga centre is also closely related to the Andean Pacific centre. The degree of relationship nevertheless diminishes with decreasing mobility. Thus species ranges comparable to those of *Fluvicola nengeta* are not known in reptiles. On the other hand in reptiles close connections with the Caribbean centre can be demonstrated (cf. *Crotalus durissus*). A few Caatinga faunal elements have also succeeded in reaching the 'islands' of campo surrounded by rain forest near Obidos and Santarem and on Marajo (fig. 56). The populations of the humming bird *Phaethornis nattereri* on these campo islands are either in gene-exchange with

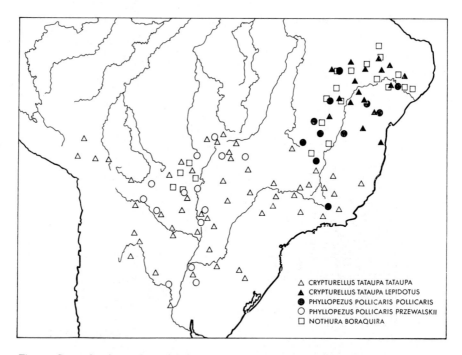

△ CRYPTURELLUS TATAUPA TATAUPA
▲ CRYPTURELLUS TATAUPA LEPIDOTUS
◉ PHYLLOPEZUS POLLICARIS POLLICARIS
○ PHYLLOPEZUS POLLICARIS PRZEWALSKII
□ NOTHURA BORAQUIRA

Fig. 54. Some Caatinga, Campo Cerrado and Chaco faunal elements. *Crypturellus tataupa lepidotus* and *Phyllopezus pollicaris pollicaris* are Caatinga faunal elements. They are replaced in the Chaco and Campo Cerrado centres respectively by the subspecies *C. tataupa tataupa* and *P. pollicaris przewalskii*. *Nothura boraquira* is a polycentric species which occurs as two disjunct populations in the Chaco centre and in the Caatinga centre.

116

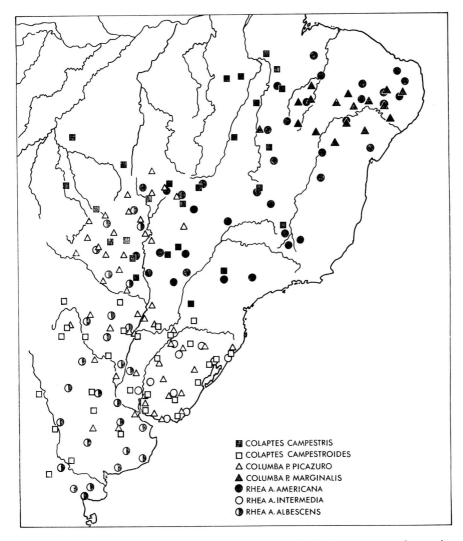

Fig. 55. Some open landscape species in South America. *Columba picazuro marginalis* must be reckoned as a Caatinga element. Its closest relative (*C. picazuro picazuro*) is polycentric (Chaco, Uruguayan and Pampa centres). *Rhea americana* is a characteristic species of the open landscapes of South America. The nominate form inhabits the Campo Cerrado and Caatinga centres. It is represented by other subspecies in the Uruguayan centre (*R. a. intermedia*) and in the Pampa and Chaco centres (*R. a. albescens*). The same holds for the campo woodpecker *Colaptes campestris* which is a Campo Cerrado element. In the Uruguayan and Pampa centres it is replaced by a vicarious species *Colaptes campestroides*.

their parent population or else can have reached the campo islands only in historical times. Most of the other populations of the campo islands that can be derived from the Caatinga centre are at least subspecifically distinct (cf. MÜLLER 1970, 1971, 1972).

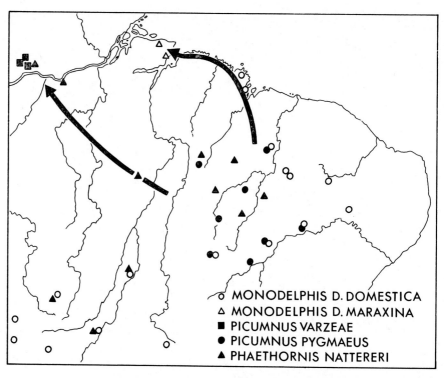

Fig. 56. Probable immigration routes (arrows) by which Caatinga faunal elements reached the hyleal campo islands of Amazonia during an arid phase (cf. fig. 97.)

The occurrence of endemic subspecies of *Gymnodactylus geckoides* and *Epichrates cenchria* in the Caatinga centre is of interest. Both species also have a subspecies in the Serra do Mar centre (fig. 57). They indicate that, given suitable pre-adaptations, particular taxa can invade very different biotopes.

The subspecies *G. g. darwini* of *Gymnodactylus geckoides* is a faunal element of the Serra do Mar. The nominate form of the species inhabits the Caatinga centre and the subspecies *G. g. amarali* the Campo Cerrado centre.

The giant snake *Epichrates cenchria* is represented in the Caatinga centre by the endemic subspecies *E. c. xerophilus*, in the forests of Espirito Santo by the endemic subspecies *E. c. hygrophilus* and in the Campo Cerrado centre by the subspecies *E. c. crassus*. *Gymnodactylus geckoides* is represented on the island São Sebastião by the same subspecies as on the mainland opposite, although the island population has been isolated for 9000 years (cf. MÜLLER 1968). The remaining faunal elements of the Caatinga centre, on the other hand, are typical indicator species of the open landscape of South America. Polycentric and polytypic species show a subspecific differentiation of range that coincides with the Caatinga and Campo Cerrado centres. Thus *Crotalus durissus* occurs as distinct endemic subspecies both in the Caatinga and the Campo Cerrado centres. It follows that both centres are closely related to each other. Changes in

118

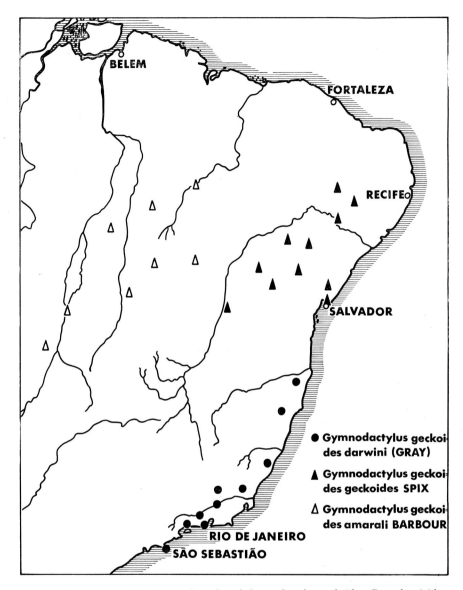

Fig. 57. Distribution of the three subspecies of *Gymnodactylus geckoides. G. g. darwini* is a Serra do Mar element, *G. g. geckoides* is a Caatinga element and *G. g. amarali* is a Campo Cerrado element.

climate would cause fluctuations in range between the Campo Cerrado and Caatinga centres on the one hand and the Chaco centre on the other. The resulting intermixture has sometimes produced a type of range which includes all these three centres, without showing any subspecific differentiation (cf. fig. 55). An animal with such a range cannot be ascribed to a centre. But the existence of this type of distribution does not argue against the separate existence of the Campo Cerrado, Chaco and Caatinga centres.

32. The Campo Cerrado centre

The centre (fig. 5, no. 32) can be defined on the ranges of *Melanopareia torquata,
Lycalopex vetulus* (fig. 58) and *Ramphocelus carbo centralis* (cf. NOVAES 1959).
Its faunal elements are adapted to the vegetational formation of the Campo
Cerrado (see SCHMITHÜSEN 1968).

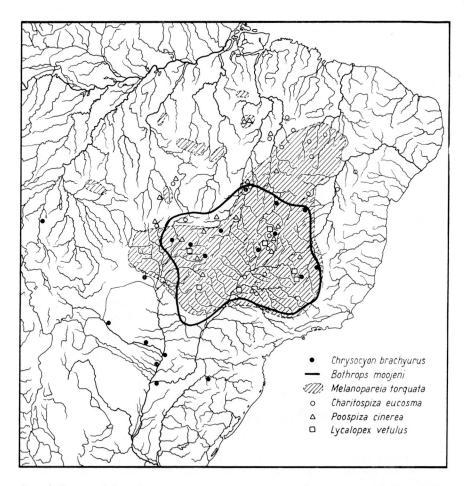

●	*Chrysocyon brachyurus*
—	*Bothrops moojeni*
▨	*Melanopareia torquata*
○	*Charitospiza eucosma*
△	*Poospiza cinerea*
□	*Lycalopex vetulus*

Fig. 58. Ranges of faunal elements of the Campo Cerrado centre. Note the occurrences of
Melanopareia torquata as populations isolated by rain forest outside the main range of the
species.

Of the 97 mammals that occur more than 23% prove to be endemic (CABRERA
1957–1962). And of the 347 bird species more than 12% are endemic, not count-
ing as such those forms which occur in the Caatinga and Chaco centres as well

as the Campo Cerrado centre (PINTO 1944, 1964, SICK 1965, MEYER DE SCHAUEN-SEE 1966, MÜLLER 1972).

The reptiles also have a number of remarkable endemics which indicate with certainty the close relationship of the Campo Cerrado centre with the Chaco and Caatinga centres. These are:

Lygophis paucidens
Lygophis lineatus dilepis
Crotalus durissus collilineatus
Chironius flavolineatus
Cnemidophorus ocellifer

Bothrops moojeni
Epicrates cenchria crassus
Hoplocercus spinosus
Leimadophis poecilogyrus intermedius

In addition there also occur in the Campo Cerrado subspecifically distinct populations of reptile species which are closest related to subspecies found in one of the Amazonian centres (Madeira or Para centre) or in the Serra do Mar centre e.g. *Constrictor constrictor amarali, Spilotes pullatus anomalepsis, Bothrops neuwiedi goyasensis.* It is striking that these endemic subspecies are all nocturnal except for *Spilotes pullatus anomalepsis* which is very eurytopic (cf. MÜLLER 1969, 1970).

Nocturnal habits are also characteristic of the amphibians of the Campo Cerrado which have recently been studied for the first time (CEI 1968, MÜLLER 1968, 1969, 1970). Contrary to SICK's view, there are indicator species of the Campo Cerrado among the Anura. These are:

Bufo paracnemis (cf. MÜLLER 1969, 1970)
B. granulosus major (cf. MÜLLER 1968)
B. ocellatus
B. rufus
Hyla venulosa (which also occurs in the flood campo of Marajó)
H. raniceps (cf. MÜLLER 1970)
H. nana
H. fuscovaria
Ceratophrys goyanus
C. ornata
Eleutherodactylus heterodactylus
Eupemphix natteri
Leptodactylus breviceps
L. chaquensis (cf. CEI 1968)
L. bufonius
L. pendatactylus labyrinthicus (cf. MÜLLER 1970)
L. gualambensis
Physalaemus cuvieri
P. centralis
Odontophrynus cultripes
Pseudopaludicola teretzi
Pleurodema verrucosa
Elachistocleis ovalis (cf. MÜLLER 1971)
Hypopachus muelleri (see MÜLLER 1971)

The closest relatives of the species of amphibia, with three exceptions, occur in the Caatinga, Uruguayan or Parana centres. The three exceptions are *Bufo paracnemis, Leptodactylus chaquensis* and *L. pentadactylus labyrinthicus* whose closest relatives live in the Amazonian forests (*B. marinus, Leptodactylus pentadactylus pentadactylus*) or in the Serra do Mar centre (*B. ictericus, L. ocellatus*).

In the birds as well, 32% of the Campo Cerrado elements have their closest relatives in a forest distribution centre. The others belong to species characteristic of the open landscapes of South America (cf. MÜLLER 1968, 1969, 1970, 1971, 1972, OLROG 1969). I shall briefly illustrate these affinities here using the families Tinamidae and Cariamidae as examples. Both these families, in MAYR's opinion, are South American in origin. In South America as a whole there are 46 species and nine genera of Tinamidae, and eight species can be reckoned as Campo Cerrado faunal elements. The Cariamidae comprises only two species placed in the separate genera *Chunga* and *Cariama*, and of these only *Cariama* with its single species *Cariama cristata* can be reckoned as a Campo Cerrado faunal element. If the distribution of the closest relatives is considered, the following picture emerges:

Campo Cerrado	Caatinga	Amazonian Centres	Serra do Mar
1. *Rhynchotus rufescens rufescens*	*R.r. catingae*	—	—
2. *Crypturellus tataupa tataupa*	*C.t. lepidotus*	—	—
3. *Nothura maculosa major*	*N.m. cearensis*	—	—
4. *Nothura boraquira*	*N. boraquira*		
5. *Nothura minor*	*N. minor*	—	—
6. *Crypturellus parvirostris*	—	—	—
7. *Crypturellus undulatus vermiculatus*	—	*C.u. adspersus*	—
8. *Crypturellus noctivagus zabele*	—	*C.n. erythropus*	*C.n. noctivagus*
9. *Cariama cristata*	—	—	—

It thus appears that of the eight Tinamid species that occur in the Campo Cerrado, only five can be considered as indicator species for open landscape, as also can *Cariama cristata*. These six forms are represented in the Campo Cerrado and Caatinga centres either by exactly the same subspecies (*Nothura boraquira, N. minor, Cariama cristata*) or else by subspecifically distinct populations. But two Campo Cerrado faunal elements i.e. *Crypturellus undulatus vermiculatus* and *C. noctivagus zabele*, have their closest relatives, not in the arid Caatinga centre, but in the centres of Serra do Mar or of the Amazonian area. The Caatinga centre is thus yet more clearly distinguished from the Campo Cerrado centre for it only shows connections with a forest centre in two single cases i.e. *Epicrates cenchria xerophilus* and *Gymnodactylus geckoides*.

However the Campo Cerrado centre has much stronger relationships with the arid biomes of the Chaco and Caatinga centres than it has with the rain forest, even though it does have a few forms adapted to the rain-forest biome. But the presence of endemics whose nearest relatives come from forest biomes shows that the Campo Cerrado has never been completely unforested. It is obviously a zone of interpenetration of grassland and forest in which, according to the prevailing climatic conditions, one or other of the two vegetational formations dominates when favoured by edaphic factors.

There is some fossil evidence concerning the history of the Campo Cerrado region available from the 'Pleistocene' of Lagoa Santa (THENIUS 1959, ROMER 1966, REIG 1968, PATTERSON & PASCUAL 1968, LANGGUTH 1969). It shows that the Campo Cerrado vegetational formation must already have existed at that time. This agrees with the occurrence in the present-day fauna of monotypic genera limited to the Campo Cerrado centre e.g. *Chrysocyon*, *Lycalopex*, which also suggests that the Campo Cerrado vegetational formation has existed for a long time. LANGGUTH was puzzled (1969) that the Canids of the Campo Cerrado i.e. *Chrysocyon* and *Lycalopex* had none of the adaptational features of the rain-forest canids such as *Atelocynus*, *Cerdocyon*, *Speothos*. But this is in no way surprising if the vegetational formation of the Campo Cerrado is very old. The canids must have been represented in South America longer than SIMPSON supposed (1950). He based his views on the fact that Canids were lacking from the Chapadmalal fauna of the late Pliocene of Argentina. However, the Pliocene faunas within the present-day geographical range of *Chrysocyon* and *Lycalopex* have not yet been studied, so his views are rather tenuously based. On the other hand it is certain that the campo fox *Dusicyon* entered South America from North America during the Pleistocene.

The occurrence of *Lycalopex*, *Dusicyon* and *Chrysocyon* all in the same beds in the Pleistocene of Minas Gerais (Lagoa Santa) is of basic importance in this connection. In the first place it indicates that at that time the region round Lagoa Santa already represented a Campo Cerrado biotope. In the second place it argues against the belief that *Chrysocyon* and *Lycalopex* evolved from *Dusicyon* species that invaded during the Pleistocene. It is much more likely that *Chrysocyon* and *Lycalopex* arrived earlier, by crossing water barriers, and that they were already separate species when *Dusicyon* reached South America.

According to results which I discuss elsewhere (MÜLLER 1970), many faunal elements of the Campo Cerrado centre have recently been in retreat, because of the destruction of the forest biome by man. But, on the other hand, although the artificial 'campo' so produced has little in common with Campo Cerrado either climatically or botanically, there are a number of Campo Cerrado faunal elements which have expanded along with the artificially expanded 'campo' e.g. *Rhynchotus rufescens rufescens*, *Leptodactylus pentadactylus labyrinthicus*. However these facts in no way show that the campo cerrado is not natural, though this was long the subject of lively controversy (HUBER 1896, 1900, 1902, KATZER 1898, 1902, IHERING 1907, WAIBEL 1921, MALME 1924, LANJOUW 1936, OTERO 1941, RAWITSCHER 1948, 1950, ROSEVEARE 1948, TROLL 1950, RAWITSCHER, HUECK, MORELLO & PAFFEN 1952, WILHELMY 1952, BEARD 1953, ALVIM 1954,

BAKKER 1954, FERRI 1954, 1955, 1960, 1963, WALTER 1954, BUDOWSKI 1956, PAFFEN 1957, HUECK 1957, ARENS, FERRI & COUTINHO 1958, KUHLMANN 1958, ANDRADE LIMA 1959, COLE 1960, FERRI & COUTINHO 1960, AZEVEDO 1962, VANZOLINI 1963, LANGE, FERRI & FERRI 1964, SICK 1965, 1966, 1968, MÜLLER 1968, 1969, 1970, 1971, 1972, SIOLI 1968, GOODLAND 1971).

33. The Serra do Mar centre

The centre (fig. 5, no. 33) lies in the rain-forest biomes of the Brazilian coastal range i.e. the Serra do Mar, broadly understood. It extends from the state of Santa Catarina in the south to the state of Pernambuco in the north, including the continental islands offshore to these states. The limits of the centre are therefore much wider than the Serra do Mar of south-east Brazil in the normal geographical sense.

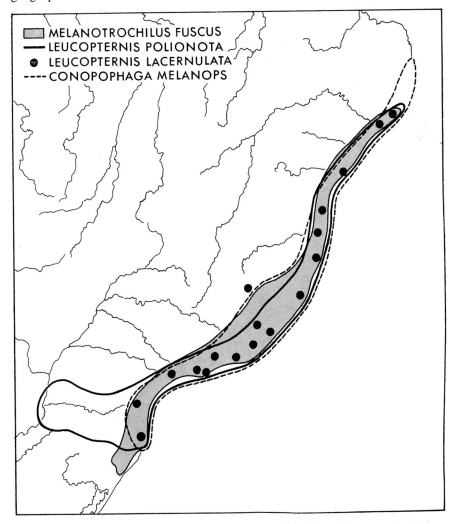

Fig. 59. Distribution of the monotypic species *Melanotrochilus fuscus, Leucopternis polionota, Leucopternis lacernulata* and *Conopophaga melanops*. These are faunal elements of the Serra do Mar centre.

Except for Florianopolis, the islands off the coast within the 50 m. isobath have a purely forest fauna. This can be regarded as indicating the long duration and constancy of the continental ranges of the species concerned, if accidental factors are carefully excluded (MÜLLER 1968, 1970). The islands vary greatly in age from 4000 to 11,000 years as can be shown not only on geomorphological and glacial eustatic grounds (BIGARELLA 1965 etc.) but also from studies in zoo-geography and evolutionary genetics (MÜLLER 1968, 1969, 1970, 1971, 1972).

Faunal elements which can certainly be reckoned to the centre mostly occur on the islands as well and show two types of geographical range: The first type

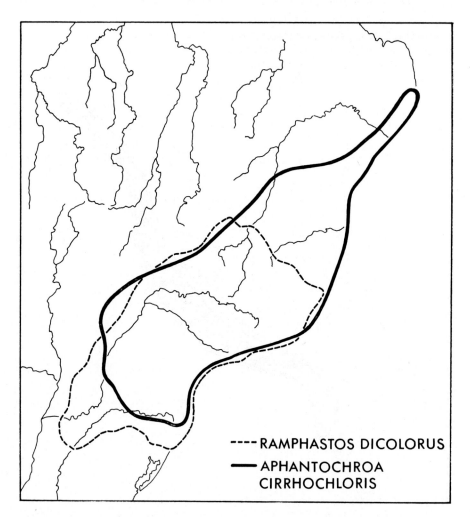

Fig. 60. Ranges of two expansive Serra do Mar faunal elements. *Ramphastos dicolorus* and *Aphantochroa cirrhochloris* extend from the forests of the Serra do Mar centre, by way of the gallery forests into the Campo Cerrado centre.

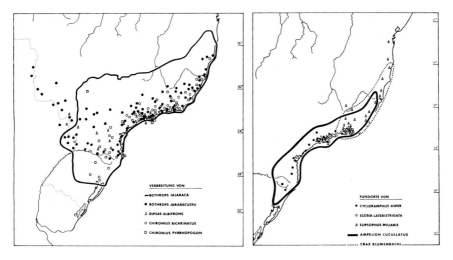

Fig. 61. Ranges of expansive Serra do Mar elements (left) and stationary elements (right) (after MÜLLER 1970).

is represented by faunal elements whose species ranges are limited to the distribution centre as defined above (cf. MÜLLER 1968). Examples are: *Melanotrochilus fuscus* and *Conopophaga melanops* (fig. 59). The second type is represented by expansive faunal elements whose western limits of range are correlated with the edges of forest in the Parana region. Examples are: *Ramphastos dicolorus*, *Aphantochroa cirrhochloris*, *Bothrops jajaraca*, *Bothrops jararacussu*, cf. fig. 60, 61. Some of these elements also penetrate into the general region of the *Araucaria* forests. Detailed study shows, however, that in such cases they only occur at rather low altitudes where, in point of fact, *Araucaria* forests are either lacking or mixed with *Podocarpus lambertii* (MÜLLER 1970). Representatives of both types of distribution strictly avoid the high campos of the Serra do Mar (cf. vegetation maps in SCHMITHÜSEN 1968) and the restinga zone of the coast.

Analysis of the endemicity of the area shows that the Serra do Mar centre has the highest absolute numbers of monotypic bird and amphibian genera of any centre. The list below further indicates that the greatest richness in endemics occurs in the bird families which MAYR (1964) considered to be originally South American and which contain an overwhelming majority of species adapted to rain forest. The list only contains families that are important for a zoogeographical analysis of the rain-forest fauna. In it: I = number of genera in South America; II = number of species in South America; III = genera occurring in the Serra do Mar centre; IV = species occurring in the Serra do Mar centre; V = endemic genera of the Serra do Mar centre; VI = endemic species in the Serra do Mar centre.

Family:	I	II	III	IV	V	VI
Galbulidae	5	15	1	1	1	1
Ramphastidae	6	41	3	5	1	2
Thraupidae	47	179	19	39	4	19
Cotingidae	35	88	16	21	4	11
Pipridae	19	52	7	11	1	5
Tyrannidae	102	302	53	79	5	20
Furnariidae	55	211	15	23	5	17
Formicariidae	51	227	17	38	15	24
Rhinocryptidae	12	28	3	4	2	4
Psittacidae	25	112	11	26	1	9
Caprimulgidae	11	27	8	10	2	2
Trochilidae	90	242	21	33	6	13
Bucconidae	10	32	5	5	0	1
Capitonidae	3	12	0	0	0	0
Picidae	14	83	9	17	0	6
Falconidae	8	23	6	11	0	0
Cuculidae	7	23	7	10	0	0
Strigidae	11	28	9	13	0	3
Cracidae	9	40	4	5	0	2
Phasianidae	4	14	1	1	0	1
Columbidae	9	48	7	12	0	1
Tinamidae	9	46	3	7	0	2
Ardeidae	17	24	11	13	0	0
Accipitridae	25	59	20	31	0	2
Apodidae	7	22	5	7	0	2
Trogonidae	2	14	1	5	0	0
Momotidae	4	4	1	1	0	0
Parulidae	13	59	2	4	0	1
Coerebidae	9	37	5	7	0	1
Fringillidae	55	193	29	32	0	9
Motacillidae	1	7	1	4	0	1
Bombycillidae	1	1	0	0	0	0
Vireonidae	4	23	3	4	0	0
Icteridae	27	60	9	11	0	1
Mimidae	5	11	1	2	0	0
Turdidae	6	33	3	7	0	0
Sylvidae	3	8	2	2	0	1
Dendrocolaptidae	13	46	4	6	0	2
Conopophagidae	2	10	2	3	0	1
Troglodytidae	9	38	4	5	0	0

Endemic bird genera of the Serra do Mar centre, which are all monotypic genera except *Mackenziaena*, *Merulaxis* and *Carpornis*, are as follows:

Triclaria malachitacea	(Psittacidae)
Macropsalis creagra	(Caprimulgidae)
Eleothreptus anomalus	(Caprimulgidae)
Ramphodon naevius	(Trochilidae)
Melanotrochilus fuscus	(Trochilidae)

Stephanoxis lalandi	(Trochilidae)
Leucochloris albicollis	(Trochilidae)
Aphantochroa cirrhochloris	(Trochilidae)
Clytolaema rubricauda	(Trochilidae)
Jacamaralcyon tridactyla	(Galbulidae)
Baillonius bailloni	(Ramphastidae)
Clibanornis dendrocolaptoides	(Furnariidae)
Oreophylax moreirae	(Furnariidae)
Anabazenops fuscus	(Furnariidae)
Cichlocolaptes leucophrys	(Furnariidae)
Heliobletus contaminatus	(Furnariidae)
Hypoedaleus guttatus	(Formicariidae)
Batara cinerea	(Formicariidae)
Mackenziaena leachii & M. severa	(Formicariidae)
Biatas nigropectus	(Formicariidae)
Rhopornis ardesiaca	(Formicariidae)
Merulaxis ater & M. stresemanni	(Rhinocryptidae)
Psilorhamphus guttatus	(Rhinocryptidae)
Phibalura flavirostris	(Cotingidae)
Tijuca atra	(Cotingidae)
Carpornis cucullatus & C. melanocephalus	(Cotingidae)
Calyptura cristata	(Cotingidae)
Ilicura militaris	(Pipridae)
Yetapa risoria	(Tyrannidae)
Muscipipra vetula	(Tyrannidae)
Ceratotriccus furcatus	(Tyrannidae)
Leptotriccus sylviolus	(Tyrannidae)
Culicivora caudacuta	(Tyrannidae)
Stephanophorus diadematus	(Thraupidae)
Orthogonys chloricterus	(Thraupidae)
Pyrrhocoma ruficeps	(Thraupidae)
Orchesticus abeillei	(Thraupidae)

Taoniscus nanus and *Donacospiza albifrons* have been left out of the list because, although they occur in the region of the distribution centre, they are not restricted to forest.

Serra do Mar faunal elements in the amphibian families Pipidae and Leptodactylidae are listed below. (E) indicates an endemic genus and (M) a monotypic genus.

Pipa carvalhoi
Basanitia bolbodactyla (E)
[LYNCH (1968) has returned *Basanitia* to the genus *Eleutherodactylus*]
B. lactea (E)
Ceratophrys appendiculata
C. boiei
C. cristiceps
C. fryi
C. renalis
C. schirchi
C. varia
Craspedoglossa sanctaecatharinae (E)
C. stejnegeri (E)

Crossodactylodes pintoi (M)
Crossodactylus aeneus (E)
C. dispar (E)
C. gaudichaudii (E)
C. schmidti (E)
C. trachystoma (E)
Cycloramphus asper (E)
C. diringshofeni (E)
C. eleutherodactylus (E)
C. fulginosus (E)
C. granulosus (E)
C. neglectus (E)
C. ohausi (E)
C. umbrinus (E)
Eleutherodactylus argyreornatus
E. hoehnei
E. nasutus
E. parvus
E. pliciferus
E. ramagii
E. venancioi
Elosia aspersa (E)
E. lateristrigata (E)
E. mertensi (E)
E. nasus (E)
E. perplicata (E)
E. pulchra (E) [above 1500 m. according to LUTZ (1952)]
Euparkerella brasiliensis (M)
Thoropa bolitoglossus
T. lutzi
T. miliaris
T. petropolitanus
T. verzus (nomen novum—GORHAM & JOHN 1966)
Holoaden bradei (E)
H. luederwaldti (E)
Leptodactylus gaigeae
L. pumilio
L. pustulatus
L. troglodytes
Macrogenioglottus aliopioi (M)
Megaelosia goeldii (M)
Paratelmatobius lutzii (E)
P. pictiventris
Phrynanodus nanus (M) [placed in *Eleutherodactylus* by LYNCH 1968]
Physalaemus maculiventris
P. moreirae
P. nanus
P. olfersi
P. signiferus
Zachaenus parvulus

In the amphibian family Hylidae at least 71 species can be reckoned as faunal elements of the Serra do Mar centre, on the basis of data in COCHRAN (1955), LUTZ (1963), LUTZ & BOKERMAN (1963), BOKERMAN (1964), CEI (1968), MÜLLER

(1968). On checking localities the Iguanids *Anisolepis grilli, Anolis nasofrontalis, Enyalius bibroni, E. iheringi* and *E. catenatus* are faunal elements of the centre as also are the Teiids *Arthroseps werneri, Anotosaura collaris, colobodactylus taunayi, Placosoma cordylinum,* and *Stenolepis ridleyi,* which both belong to monotypic genera. The same holds for the Crotalids *Bothrops megaera, B. jajaraca, B. jararacussu, B. insularis, B. pradoi* and *B. pirajai.*

The Serra do Mar centre can be divided easily into three subcentres. The northernmost lies between Salvador (= Bahia) and Recife and can be called the **Pernambuco subcentre** (fig. 62). Its position can be defined by the limits of range of *Bothrops megaera* and *Tinamus solitarius pernambucensis.*

The second subcentre, which is adjacent to the Pernambuco subcentre and south of it, lies in the rain-forest biome of Ilheus and can be called the **Bahia**

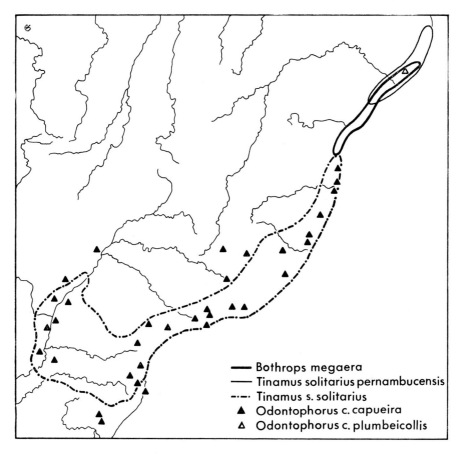

Fig. 62. Distributions in the Serra do Mar centre. *Bothrops megaera, Odontophorus capueira plumbeicollis* and *Tinamus solitarius pernambucensis* are faunal elements of the Pernambuco subcentre of the Serra do Mar centre. Their closest relatives inhabit the Bahia and Paulista subcentres of this centre.

subcentre. Expansive faunal elements of this centre occur as far north as the Rio Paraguassu. Species with good flying ability such as *Cichlolaptes leucophrys* (fig. 63) and *Attila rufus* can penetrate the Pernambuco subcentre if they meet no ecological competitors. Bahia faunal elements reach their southern limits south of Victoria (Estado de Espirito Santo e.g. *Bothrops pradoi*). In a few cases they even occur in the region of the Pico de Bandeira and at the northern margin of the Serra de Orgãos.

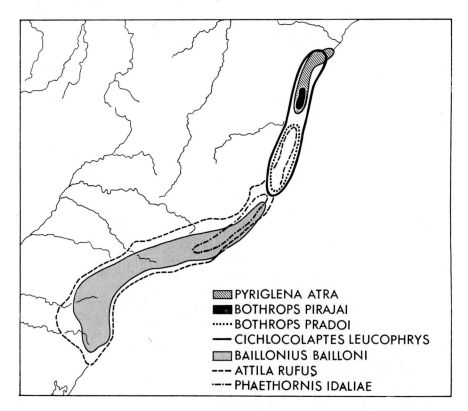

PYRIGLENA ATRA
BOTHROPS PIRAJAI
BOTHROPS PRADOI
CICHLOCOLAPTES LEUCOPHRYS
BAILLONIUS BAILLONI
ATTILA RUFUS
PHAETHORNIS IDALIAE

Fig. 63. Distributions in the Serra do Mar centre. *Pyriglena atra, Bothrops pirajai, Bothrops pradoi* and *Cichlocolaptes leucophrys* are faunal elements of the Bahia subcentre of the Serra do Mar centre. *Attila rufus, Phaethornis idaliae* and *Baillonius bailloni* must be considered as elements of the Paulista subcentre, although they also exist in the Iguaçu area.

The third and most important of the subcentres is the **Paulista subcentre.** It lies in the south-eastern part of the Brazilian coast range from Florianopolis in the south to Cabo Frio in the north, and can be defined by the limits of range of the Ramphastid *Baillonius bailloni* which belongs to a monotypic genus (fig. 63). Highly mobile species link all three subcentres directly together (fig. 62).

The strong orographic subdivision which exists increases the degree of isolation of Serra do Mar populations and favours the appearance of local centres

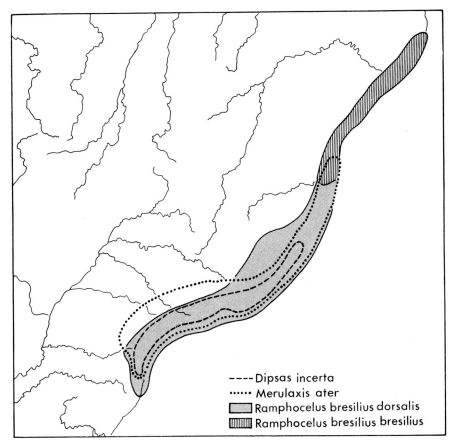

Fig. 64. Distributions in the Serra do Mar centre. *Dipsas incerta* and *Merulaxis ater* are further examples of the type of distribution associated with faunal elements of the Paulista subcentre of the Serra do Mar centre. The range of *Rhamphocelus bresilius* is polycentric but its subspecies are monocentric faunal elements. Thus *R. b. dorsalis* is an element of the Paulista subcentre and *R. b. bresilius* is an element of the Pernambuco subcentre.

of subspeciation. Neither this orographic effect however, nor present-day climatic conditions, will explain the division of the Serra do Mar centre into three. But if the ecological valency of the species and subspecies occurring in the individual subcentres is considered, then three factors emerge as important barriers to distribution. They are: 1. the 1400–1500 m. contour (fig. 65); 2. unforested landscapes; and 3. water barriers.

In the region of Cabo Frio there are broad areas of restinga which have probably acted as barriers preventing Bahia elements moving south or Paulista elements moving north. It appears however, that expansive elements ascribed to both subcentres have circumvented this campo barrier on its landward side by using the lower parts of forested mountains (cf. Discussion). However, this by-pass must obviously have been interrupted in the past.

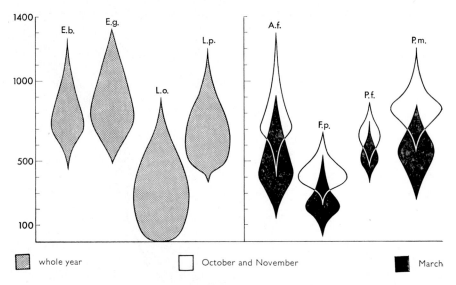

Fig. 65. Vertical distribution on the island of São Sebastião of four amphibia adapted to rain forest and four species of birds. The amphibia are *Eleutherodactylus binotatus, E. guntheri, Leptodactylus ocellatus* and *L. pentadactylus*. The birds are *Amazona farinosa, Forpus passerinus, Pyrrhura frontalis* and *Pionus maximiliani*. The seasonal change in distribution of the birds is correlated with the breeding season in October and November. Based on observations lasting several months in 1964–1965, 1967 and 1969.

The closest relationships of Serra do Mar faunal elements are with the centres of the Amazonian region. Among forms strictly adapted to the rain-forest biome which occur in the Amazonian centres and in the Serra do Mar subcentres there is a striking disjunction of range. This is correlated with the extent of the Campo Cerrado. Even for species with strong flying ability it can be shown that the populations separated by this disjunction are no longer in gene-exchange with each other. Thus among birds there are 30 non-passeriform and 67 passeriform species with disjunct ranges of this type. With only three exceptions all have subspecifically distinct populations in the Serra do Mar and the Amazonian parts of their ranges. The exceptions are *Nyctibius grandis, Hylocharis sapphirina* and *Ornithion inerme*, though for this last species the occurrence at Bahia is not yet certain.

An identical type of distribution exists as between species of the same genus. In these cases species of typical rain-forest genera are endemic to the big rain-forest centres and are lacking completely in the Campo Cerrado (fig. 66, 67, 68, 69). These various levels of differentiation show that the distribution centre of the Serra do Mar has acted as a refuge effectively isolated from other forest centres at several different periods. The endemic subspecies of the Serra do Mar are all easy to diagnose.

The same type of disjunct distribution occurs in amphibia and reptiles (cf. WILLIAMS & VANZOLINI 1966, MÜLLER 1968, 1970 and fig. 70).

134

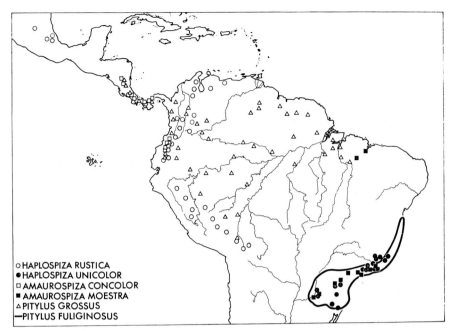

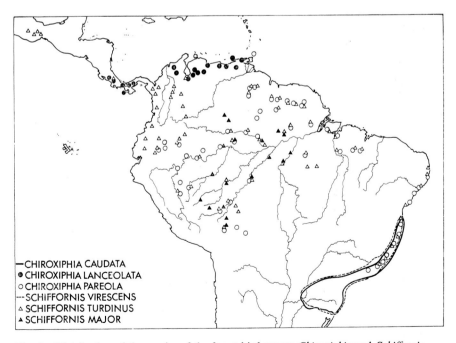

Fig. 66. Distribution of the bird genera *Haplospiza*, *Amaurospiza* and *Pitylus*. *Pitylus fuliginosus* is a Serra do Mar faunal element. Its closest relative is *P. grossus* which is a polycentric species with distinct subspecies in the Amazonian centres in the Choco centre and in the Costa Rican centre.

Fig. 67. Distribution of the species of the forest bird genera *Chiroxiphia* and *Schiffornis*.

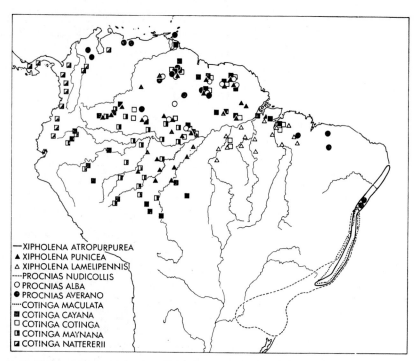

Legend (Fig. 68):

—XIPHOLENA ATROPURPUREA
▲ XIPHOLENA PUNICEA
△ XIPHOLENA LAMELIPENNIS
----PROCNIAS NUDICOLLIS
○ PROCNIAS ALBA
● PROCNIAS AVERANO
·····COTINGA MACULATA
◼ COTINGA CAYANA
□ COTINGA COTINGA
◫ COTINGA MAYNANA
◪ COTINGA NATTERERII

Fig. 68. Distribution of the species of the forest bird genera *Xipholena*, *Procnias* and *Cotinga*.

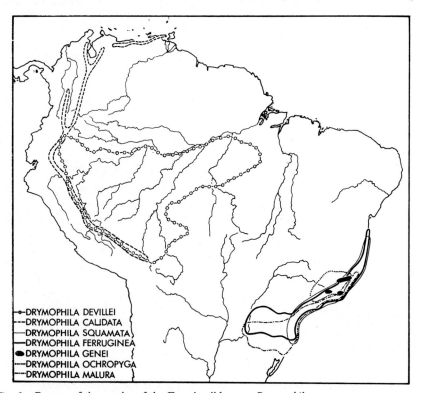

—○—DRYMOPHILA DEVILLEI
---DRYMOPHILA CALIDATA
—DRYMOPHILA SQUAMATA
—DRYMOPHILA FERRUGINEA
●DRYMOPHILA GENEI
·····DRYMOPHILA OCHROPYGA
·····DRYMOPHILA MALURA

Fig. 69. Ranges of the species of the Formicariid genus *Drymophila*.

Thus *Hyla faber* is a faunal element of the Serra do Mar whose nearest related species—*Hyla boans*—lives in the forests of Amazonia. From *Hyla boans* it is possible to derive *Hyla rosenbergi* which is a faunal element of the Choco sub-centre (Colombian Pacific centre).

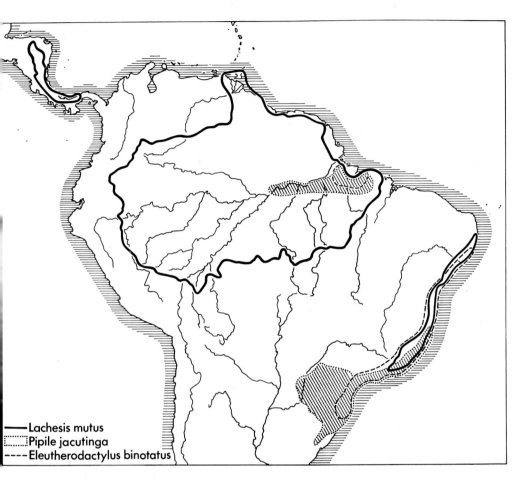

Fig. 70. Distribution of the Crotalid *Lachesis mutus*, of the Leptodactylid *Eleutherodactylus binotatus* and of the Cracid *Pipile jacutinga*.

34. The Parana centre

The centre (fig. 5, no. 34) can be defined on the ranges of *Bothrops cotiara*, *Leptasthenura setaria*, *Cinclodes pabsti* and *Amphisbaena mertensi* (fig. 71). There seems to be no direct ecological relation between the vertebrate faunal elements of the Parana centre and *Araucaria angustifolia*, unlike the situation with some invertebrates (KUSCHEL 1960, RÜHM 1969, MÜLLER 1970).

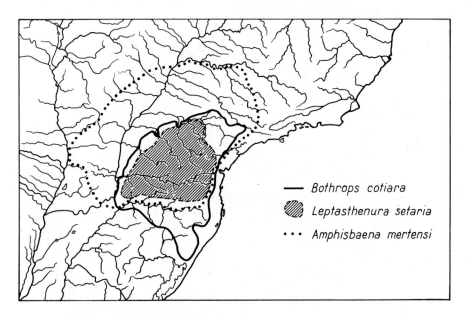

Fig. 71. Ranges of typical Parana faunal elements.

The species are endemic to the high campos of the area. The occurrence of an endemic *Cinclodes* species (cf. SICK 1969) indicates that the part of South America of definitely tropical character has now been left behind. I have written elsewhere on the ecology of the area (MÜLLER 1970). The most closely related species of *Cinclodes* occur in southern Rio Grande do Sul, Uruguay, Patagonia and the Andes (cf. Puna and North Andean centres). They have a similar distribution to the genus *Leptasthenura*. Within the Leptodactylidae only *Zachaenus sawayae* can be reckoned as a faunal element of the Parana centre (cf. COCHRAN 1953, p. 111). Its closest relative, *Z. roseus*, is a Patagonian faunal element.

The occurrence of *Araucaria angustifolia* together with the Parana faunal elements indicates only that both prefer to live in this relatively high-lying area above 800 m. (cf. AUBRÉVILLE 1948). There is no indication that the origin of the Parana faunal elements is directly connected with the origin of the *Araucaria* forests.

The centre is entirely surrounded by rain forest. It can be deduced that during the post-glacial arid phase the rain-forest biome in Rio Grande do Sul was much more strongly penetrated by campos than it now is. This is shown by studies of the geology (DELANEY 1963, HURT 1964, AB'SABER 1965) and plant sociology (RAMBO 1948, 1956) and also by zoogeographical studies that I carried out in 1967 and 1969 on the island of Florianopolis and in the region of Rio Grande do Sul (MÜLLER 1970).

The warming-up of the high campos that occurred at the same time certainly led to the invasion of a number of lowland species into the Parana distribution centre e.g. *Bothrops alternatus, Crotalus durissus* cf. MÜLLER (1970). But at this period the Parana faunal elements more strongly adapted to a cool climate could scarcely have been in an expansive state. The fact that in most cases they are specifically distinct, except for *Crotalus durissus terrificus* which is only subspecifically distinct, suggests a long period of isolation. The few lowland species that occur in the area, on the other hand, are at most subspecifically distinct, e.g. *Bothrops neuwiedi paranaensis*. Furthermore, even within the centre they show a strong preference for low-lying situations (MÜLLER 1970).

35. The Uruguayan centre

The position of this centre (fig. 5, no. 35), can be defined on the ranges of its faunal elements: *Anopsibaena kingii, Pleurodema darwinii, Cthonerpeton in-distinctum, Limnornis curvirostris* and *Anisolepis undulatus* (fig. 72).

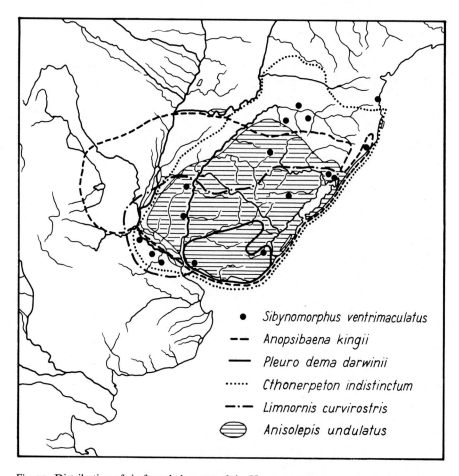

Fig. 72. Distribution of six faunal elements of the Uruguayan centre.

For faunal elements that are not very vagile the Rio Parana forms the western and southern limit of the centre while the rain-forest biome of Rio Grande do Sul and Santa Catarina forms the northern limit. A zoogeographical analysis of the herpetofauna of the island of Florianopolis showed that 20% of the species could be reckoned as Uruguayan faunal elements (fig. 73, see MÜLLER 1970). This cannot be explained from the present-day ecology since the island

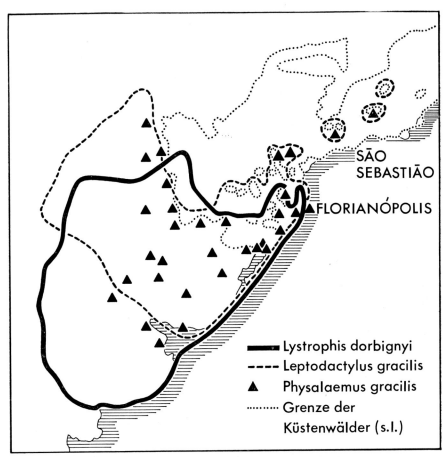

■■■■ Lystrophis dorbignyi

---- Leptodactylus gracilis

▲ Physalaemus gracilis

········· Grenze der
　　　　　Küstenwälder (s.l.)

Fig. 73. Ranges of three faunal elements of the Uruguayan centre. Note the isolated occurrences in the forests of the Serra do Mar and on the island of Florianopolis (after MÜLLER 1970). Limits of coastal forest (s.l.) dotted.

is screened off to the south from the Uruguayan centre by the rain-forest biome of the mainlands. It can only be supposed that expansive Uruguayan elements reached the island at a time when no such forest barrier existed and when the island was also in contact with the mainland (cf. Discussion).

The low degree of differentiation of the island populations shows that these conditions must have existed during post-glacial times. Thus some of them show a weak subspecific differentiation while on the island of Campeche, which lies east of Florianopolis, there is an easily diagnosed subspecies of *Leptodactylus gracilis* (cf. MÜLLER 1968). The expansive Uruguayan elements did not reach only the island of Florianopolis, but have also been found in the isolated high campos within the rain forest of the Serra do Mar e.g. *Leptodactylus gracilis*, *Physalaemus gracilis* cf. fig. 73, MÜLLER (1970). This shows that the results got from Florianopolis also apply to the mainland.

It therefore appears that during the post-glacial arid phase the Uruguayan centre expanded towards the north. The evidence for this does not come only from zoogeography but also from recent geological studies (AB'SABER 1962, 1965, HURT 1964, BIGARELLA 1965), from geography (WILHELMY 1952, 1957, AB'SABER 1965) and from climatology (REINKE 1962) cf. Discussion.

370 species of birds have been recorded in Uruguay on the basis of information in BARROWS (1884), APLIN (1894), TREMOLERAS (1920, 1927), DEVINCENZI (1925–1928), DARBENE (1926), WETMORE (1926), ALVAREZ (1933, 1934), TEAGUE (1955), BARATTINI & ESCALANTE (1958), CUELLO (1959), ESCALANTE (1959, 1960), CUELLO & GERZENSTEIN (1962) and SACCONE (1962). Of these species the following can be reckoned as Uruguayan faunal elements, where (M) = monotypic genus:

Rhea americana intermedia	*Yetapa risoria* (M)
Chauna torquata	*Molothrus rufoaxillaris*
Anodorhynchus glaucus	*Xanthopsar flavus* (M)
Limnornis curvirostris (M)	*Amblyramphus holosericeus* (M)
Limnoctites rectirostris (M)	*Pezites defilippii*
Leptasthenura platensis	*Cyanoloxia glaucocaerulea*
Cranioleuca sulphurifera	*Sporophila palustris*
Asthenes baeri	*Poospiza nigrorufa nigrorufa*
A. hudsoni	*Paroaria coronata*
Spartanoica maluroides (M)	*Gubernatrix cristata*
Xolmis dominicana	

Two Crotalid, three Iguanids and one Anguid species are likewise faunal elements of the Uruguayan centre i.e. *Bothrops alternatus* (Crotalidae) cf. range in MÜLLER 1970, *Bothrops neuwiedi pubescens* (Crotalidae), *Anisolepis undulatus* (Iguanidae), *Leiosaurus paronae* (Iguanidae), *Proctotretus azureus* (Iguanidae), *Ophiodes vertebralis* (Anguidae).

In the amphibian family Leptodactylidae ten species prove to be faunal elements of the Uruguayan centre. These are:

Ceratophrys bigibbosa	*Physalaemus gracilis*
Leptodactylus gracilis	*P. henseli*
L. prognathus	*P. riograndensis*
Limnomedusa macroglossa	*Pleurodema darwinii*
Odontophrynus americanus	*Pseudopaludicola falcipes*

Faunal elements of the Serra do Mar centre penetrate into the Uruguayan centre along gallery forests (OREJAS MIRANDA 1958, MÜLLER 1968, 1970). These populations of the Serra do Mar centre in Uruguay are nevertheless not sub-specifically distinct (DEVINCENZI 1935, ACOSTA Y LARA 1950, CUELLO & GERZEN-STEIN 1962, MÜLLER 1970).

36. The Chaco centre

The centre (fig. 5, no. 36) can be defined on the ranges of *Nothura chacoensis* and *Chunga burmeisteri* (fig. 53). Zoogeographically the Chaco area of Argentina and Paraguay is very complex and different authors have reached contradictory conclusions about it. The reasons for this are twofold. Firstly, this is an area where the ranges of faunal elements of very different origins overlap (BOETTGER 1885, BOULENGER 1894, 1898, PERACCA 1895, MEHELY 1904, BERTONI 1914, 1928, 1939, KRIEG 1927, 1931, 1936, LAUBMANN 1930, 1939–1940, EISENTRAUT 1931, 1935, EMDEN 1935, BRODKORB 1937, KANTER 1936, MÜLLER & HELLMICH 1936, SHREVE 1948, HELLMICH 1960, CEI 1950, CEI & BERTINI 1961, STEINBACHER 1962). Secondly, forms adapted to the rain forest have penetrated far to the south inside gallery forests e.g. *Iguana iguana, Anolis chrysolepis* (cf. MÜLLER 1970). Because of this complexity individual authors have included very different zoogeographical entities as faunal elements of the region. However, by considering first those species and subspecies whose limits of range coincide with the Chaco centre, the interpretation of this region as an independent dispersal centre becomes quite obvious.

Of the 268 species of birds mentioned by STEINBACHER (1962, 1968) as occurring in West Paraguay, 21 can be reckoned as endemic to the Chaco centre. These are:

Nothura chacoensis
Chunga burmeisteri
Ortalis canicollis ungeri
Myiopsitta monacha cotorra
Strix rufipes chacoensis
Chrysoptilus melanochloros nigroviridis
Picumnus cirratus pilcomayensis
Lepidocolaptes angustirostris angustirostris
Campylorhamphus trochilirostris lafresnayanus
Thamnophilus caerulescens paraguayensis
Asthenes baeri chacoensis
Myiarchus swainsoni ferocior
Elaenia spectabilis spectabilis
Sublegatus modestus brevirostris
Camptostoma obsoletum obsoletum
Spinus magellanicus alleni (cf. Roraima centre)
Zonotrichia capensis mellea
Coryphospingus cucullatus fargoi
Embernagra platensis olivescens
Thraupis sayaca obscura
T. bonariensis schulzei

Species which have the Paraguay river as a subspecies limit are adapted more to the arid forest than to open grassland. Four rodent species should be reckoned as Chaco faunal elements. These are: *Akodon chacoensis, Scapteromys chacoensis, Pseudoryzomys waurini, Ctenomys (Chacomys) conoveri.*

Reptiles are represented by two indicator species in the arid western Chaco i.e. *Tropidurus melanopleurus* and *Liolaemus chacoensis*. The phyletic affinities of the *Tropidurus* of the Chaco area with those of the Andean Pacific centre need to be re-examined using a very large quantity of material. HELLMICH (1960) considered that 6% of the reptiles of the western Chaco were endemic. This figure is certainly too high, however, since the included *Leposternon* species need extensive revision (GANS, 1967, Bull. Amer. Nus. Nat. Hist. 135).

As regards their adaptational mechanisms, Chaco faunal elements show no obvious specialisation for life in an arid region. This suggests that the present-day aspect of the western Chaco is relatively young. This supposition is supported by geological and geomorphological investigations (ECKEL 1952, KANTER 1936, WILHELMY 1954, CASTELLANOS 1959 in HELLMICH 1960). These show that in this region there were four pluvial phases in the Pleistocene period supposedly synchronous with the glacial phases of the northern hemisphere. Towards the end of the deposition of the Pampaean Formation there were big lakes in the Chaco ('Lughanese' of Ameghino). This was followed by an arid cycle with the formation of saline lagoons, marking the end of the Pleistocene (Platense Formation). The subsequent 'Piso Nonense' led to a damper climate for a short time, followed by another arid phase ('Piso Cordobense') characterised by loess storms. This was followed by a damp period which lasted into the sixteenth century. From this time onward there has been another increase in aridity. The basin of the Rio Paraguay is said not to have undergone any essential displacement during the Pleistocene (ALMEIDA 1945, PUTZER 1958).

A number of Chaco faunal elements, such as *Tupinambis rufescens*, appear to be expanding their ranges. The explanation, however, as KRIEG (1936) and HELLMICH (1960) proposed, is probably that certain forms, because of increased aridity, are retreating outwards from the Chaco region into more favourable situations around it.

This explanation agrees with the fact that amphibians are very weakly represented in the Chaco with only one endemic species (*Chacophrys*, monotypic genus). It is true that FREIBERG (1942), VELLARD (1946) and CEI (1955) list many amphibia as 'typical Chaco elements'. These include:

Bufo paracnemis	*L. mystaceus*
B. granulosus dorbignyi	*L. gracilis*
Bufo arenarum	*L. laticeps*
Ceratophrys ornata	*Physalaemus fuscomaculatus*
C. pierottii	*P. cuvieri*
Odontophrynus americanus	*Pseudis paradoxa*
Odontophrynus cultripes	*P. limellum*
Lepidobatrachus asper	*Pseudopaludicola falcipes*
Leptodactylus podicipinus	*Hyla phrynoderma*
L. ocellatus	*H. nana*
L. chaquensis	*H. lindneri*
L. bufonius	*H. faber*
L. prognathus	*H. venulosa*
L. sybilator	*H. nasica*

H. raddiana	Dendrophryniscus stelneri stelneri
H. spegazzinii	D. rubiventris
Phyllomedusa sauvagii	Elachistocleis ovalis bicolor
P. hypochondrialis	Hypopachus muelleri

However, if the geographical ranges of these forms are studied, they are not Chaco elements in the sense used here. One of the forms listed i.e. *Hyla faber* is actually lacking in the Chaco; the specimen mentioned by CEI (1955) is in fact an unpatterned female of *Hyla venulosa* (see MÜLLER 1968). The others are either indicator species of the campos of central South America e.g. *Bufo paracnemis, Hypopachus muelleri, Elachistocleis ovalis, Physalaemus fuscomaculatus* (MÜLLER 1969, 1970), or else they are frogs that live the whole year round in water (*Pseudis paradoxa, P. limellum*—MÜLLER 1970).

The north-east limits of the centre seem to have fluctuated greatly (MÜLLER 1970). In the Pantanal of the Mato Grosso the influence of Chaco faunal elements has already greatly decreased. The northern Pantanal, between the Rio Piquiri and Rio Taquari does not form part of the true Chaco centre. The vegetation of this region in sandy elevated situations agrees to a very large extent with that of Cuiaba, Rondonopolis and Poxoeu. The species and subspecies which occur here are either Campo Cerrado faunal elements e.g. *Bothrops moojeni, Hoplocercus spinosus*, or Uruguayan or Pampa faunal elements e.g. *Lystrophis semicinctus, Rhea americana albescens.*

The Chaco centre has strong affinities with the Brazilian Caatinga centre (cf. subspecies ranges of *Crypturellus tataupa* and *Phyllopezus pollicaris* in fig. 54). Affinities with the Andean Pacific centre have already been discussed under that centre (cf. fig. 42).

37. The Monte centre

This centre (fig. 5, no. 37) can be defined on the ranges of: *Batrachophrynus patagonicus, Lepidobatrachus salinicola, Pleurodema nebulosa, Odontophrynus occidentalis, Chlamyphorus truncatus, Bothrops ammodytoides, Cnemidophorus longicaudus, Homonota whitii* (fig. 74) and *Leiosaurus catamarcensis* (see GALLARDO 1960). Its faunal elements are most closely related to those of the Pampa and Chaco centres. Among the rodents, however, there are faunal elements of the Monte centre whose phyletic relationships are still not clear (*Octomys mimax, Tympanoctomys barrerae*). Some Patagonian faunal elements penetrate the Monte centre from the south and indeed some of them pass right through the Monte centre into the Puna centre of the Andes (cf. the ranges of *Phoenicopterus chilensis* and *Pterocnemia pennata*).

The Pliocene fauna of the centre is particularly well known (PATTERSON & PASCUAL 1968), while faunas of Miocene or Pleistocene age have only been found in a few places. The few Pre-Pliocene occurrences demonstrate that the region at that time had a mixed fauna of forest and pampa types (see also MENENDEZ 1969). At the beginning of the Pliocene (Chasicoan) the forest fauna greatly decreased, and at the end of the Pliocene (Montehermosan) it was entirely replaced by an open landscape fauna. No comparable fossil occurrences have yet been found west of the Andes so that, on direct evidence, nothing can be said of the character of the fauna there. It is quite likely that in the Pre-Pliocene it was identical with the fauna east of the present Andes but this is based only on our knowledge of the timing of the Andean orogeny. The elevation of the Andes would have been correlated with an abrupt change in the vegetational relationships on the two sides of the Andes. The faunal change in the Pliocene suggests that gene exchange between the *Nothofagus* centre and the centres of Amazonia and the Serra do Mar was interrupted when the change occurred and afterwards was only possible by a sort of 'island hopping'.

At the present day, the Monte centre is delimited by the Sierra de Cordoba to the east, the Andes to the west, the Rio Salado to the north and the Rio Colorado to the south. Expansive faunal elements show a spread in range along the margin of the montane forest biome as far as the Rio Pilcomayo in the north. Monte faunal elements do not cross the Rio Negro to the south.

The ecology of the area has been reported on exhaustively by BURMEISTER (1891), KOSLOWSKY (1895, 1896, 1898), WOLFHÜGEL (1929), MÜLLER & HELLMICH (1938, 1939), KRIEG (1939), HELLMICH (1950) and WERNER (1972). According to the data of OLROG (1963) and HOY (1971) at least 29 species of birds should be considered as endemic to the centre.

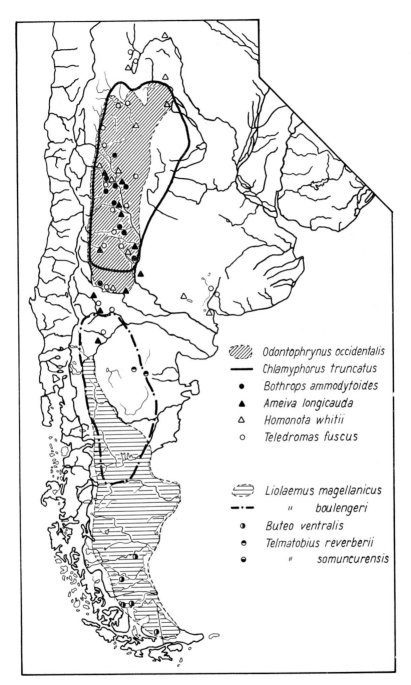

Legend within figure:

Odontophrynus occidentalis
Chlamyphorus truncatus
Bothrops ammodytoides
Ameiva longicauda
Homonota whitii
Teledromas fuscus

Liolaemus magellanicus
  "   boulengeri
Buteo ventralis
Telmatobius reverberii
  "   somuncurensis

Fig. 74. Faunal elements of the Monte centre and Patagonian centre. Monte faunal elements
are: *Odontophrynus occidentalis, Chlamyphorus truncatus, Bothrops ammodytoides, Cnemidophorus longicaudus, Homonota whitii, Teledromas fuscus.* Patagonian faunal elements are: *Liolaemus magellanicus, Liolaemus boulengeri, Buteo ventralis, Telmatobius reverberii, Telmatobius
somuncurensis.*

38. The Pampa centre

The position of the centre (fig. 5, no. 38) can be defined on the ranges of: *Calomys gracilipes*, *Chaetophractus villosus*, *Proctotretus pectinatus*, *Leptodactylus ocellatus bonariensis* and *Bufo arenarum* (fig. 75). The toad *Bufo arenarum* has been in the Pampa centre since the Pleistocene (cf. TIHEN 1962). For stationary faunal elements the Rio Salado and Rio Parana form the northern limit of the centre, while the Sierra de Cordoba forms the western limit and the Rio Negro the southern.

Species with restricted powers of movement that occur both west and east of the Sierra de Cordoba are often subspecifically distinct in the two areas (cf. *Zaedyus pichiy pichiy* and *Z. p. caurinus* in fig. 75). The same holds for populations north and south of the Rio Parana and Rio Salado, where the more northerly population belongs to the Uruguayan centre, and also north and south of the Rio Negro, where the more southerly population belongs to the Patagonian centre. Thus the Canid *Dusicyon gymnocercus*, which is adapted to open landscapes, occurs as the subspecies *D. g. antiquus* in the Pampa centre, while the second subspecies of this species i.e. *D. g. gymnocercus* can be reckoned as a Uruguayan faunal element. In the rodent *Akodon obscurus* the subspecies *A. o. benefactus* is a Pampa faunal element; in the Chaco centre the species is represented by the subspecies *A. o. lenguarum* and in the Uruguayan centre by *Akodon obscurus obscurus* itself (CABRERA 1957–1961). *Calomys laucha laucha* can likewise be considered as a Pampa faunal element; the closest related subspecies, *C. l. musculinus*, represents the species in the Monte centre. Another rodent that occurs in the open landscape of the pampa is *Calomys dubius bonariensis*, which in the Uruguayan centre is replaced by *C. d. bimaculatus*. The deer *Ozotoceros bezoarcticus* can also be mentioned here; in the Pampa centre it is represented by *O. b. celer*, in the Chaco centre by *O. b. leucogaster* and in the Campo Cerrado centre by *O. b. bezoarcticus*.

The Uruguayan and Pampa centres have many things in common. Nevertheless it is striking that if a faunal element can be ascribed without doubt to one of the two centres, in most cases it only occurs in adjacent marginal areas of the other centre. Populations in such adjacent marginal areas are not differentiated from the parent population which shows that they have only expanded into these areas very recently. The same can be said for populations of Pampa elements south of the Rio Negro. These have pushed southwards from the Pampa centre, particularly along the coast.

The endemic fauna of the centre is adapted to an unforested landscape (AZARA 1847, DARWIN 1852, BURMEISTER 1861, BERG 1898, KOSLOWSKY 1898, HARTERT & VENTURI 1909, HUDSON 1920, WETMORE 1926, FERNANDEZ 1927, PEREYRA 1937, 1951, FREIBERG 1942, ABALOS 1949, OLROG 1955, 1963, 1969, SZIDAT 1955, CEI 1956, ABALOS & BAEZ 1963, ABALOS, BAEZ & NADER 1964, BARRIO 1964, 1965). This can be shown for the birds as well as for other animals. Thus OLROG lists 902 species of birds for Argentinia of which 240 are restricted to the extreme

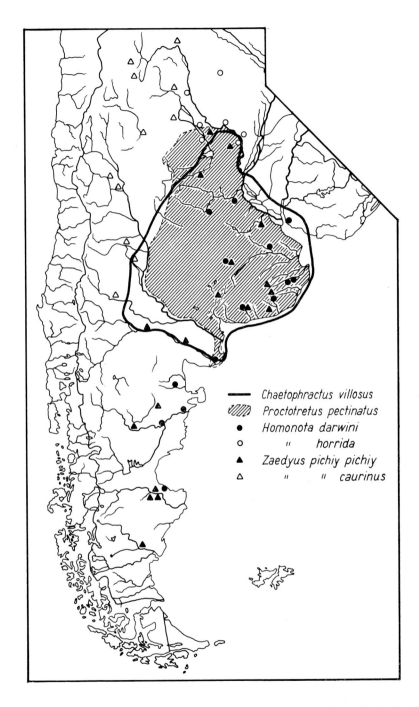

Fig. 75. Distribution of faunal elements of the Pampa centre. *Zaedychus pichiy pichiy* is a Pampa faunal element which is replaced in the Monte centre by *Z. pichiy caurinus*. The Pampa species *Homonota darwini* is replaced by the vicarious species *H. horrida* in the Chaco centre.

north of the country (Misiones etc.). The 'zona Chacopampeana', which naturally comprises both the chaco and the pampa, is inhabited by 372 species. Most of
these are widespread in the unforested areas of South America, but 52 of them
can be reckoned as faunal elements of the Pampa centre.

39. The Patagonian centre

The centre (fig. 5, no. 39) can be defined on the ranges of: *Liolaemus kingii*, *Liolaemas magellanicus* (fig. 74), *Tinamotis ingoufi* (fig. 41) and *Dusicyon griseus griseus* (fig. 76). Faunal elements of the centre also reached the Falkland Islands. This group of islands must have been to a large extent free of ice at least during the last glacial phase; this is indicated by pollen analytical data (BOYSON 1924,

Fig. 76. Distribution of *Dusicyon australis* and of the subspecies of *Dusicyon griseus* and of *Lyncodon patagonicus*. *L. patagonicus patagonicus* and *Dusicyon griseus griseus* are Patagonian faunal elements. *L. p. thomasi* and *D. g. gracilis* are faunal elements of the Monte centre. *D. g. domeykoanus* occurs in the Chilean subcentre of the Andean Pacific centre and *D. g. maullinicus* occurs in the *Nothofagus* centre.

151

DONAT 1931, AUER 1933, 1960, SKOTTSBERG 1942), by the occurrence of endemic Falkland species such as *Dusicyon australis* and by the avifauna (BENNETT 1926, 1931, OLROG 1963, HUMPHREY, BRIDGE, REYNOLDS & PETERSON 1970). Thus 11 subspecifically distinct Patagonian faunal elements among the birds have their closest relatives in the Falkland Islands, but all of them are placed in distinct subspecies.

Patagonian faunal element:	Falkland subspecies:
Podiceps rolland chilensis	*P. r. rolland*
Milvago chimango temucoensis	*M. ch. chimango*
Asio flammeus suinda	*A. f. sanfordi*
Cinclodes antarcticus maculirostris	*C. a. antarcticus*
Muscisaxicola macloviana mentalis	*M. m. macloviana*
Cistothorus platensis hornensis	*C. p. falklandicus*
Troglodytes aedon chilensis	*T. a. cobbi*
Turdus falcklandii magellanicus	*T. f. falcklandii*
Pezites militaris militaris	*P. m. falklandicus*
Melanodera melanodera princetoniana	*M. m. melanodera*
Nycticorax nycticorax obscurus	*N. n. falklandicus*

The occurrence in the Falklands of *Dusicyon australis*, which can be derived from the mainland Canid *D. griseus*, indicates that the islands had a connection by land with the mainland during the phase of maximum glaciation. The same thing is indicated by the palaeontological results of FRAY & EWING (1963). The *Dusicyon* group only reached South America in the Pleistocene.

The position of the Patagonian centre largely agrees with the extent of the 'Patagonische Kaltsteppe' of MANN (1968; see also ERIKSEN 1972). The faunal elements that belong to it are adapted to rather cold climates. It is therefore not surprising that they have their closest affinities with the Puna centre of the Andes (fig. 41). Andean forms, like *Cinclodes fuscus* and *Vultur gryphus* among others, occur in the Patagonian centre right down to sea level (fig. 77). The same is true of the Andean frog genus *Telmatobius*. Two endemic species of the Somuncura mountains i.e. *T. reverberii* and *T. somuncurensis*, have been described only recently by CEI (1969) (see fig. 74).

Tinamotis ingoufi has its closest relative (*T. pentlandii*) in the Puna centre. The same is true for Darwin's Rhea (*Pterocnemia pennata*). The nominate form of this inhabits the Patagonian centre while the subspecies *P. p. garleppi* must be reckoned as a Puna faunal element (fig. 41).

The relationship between the various centres of Argentina and Chile is indicated by the distribution of the subspecies of *Dusicyon griseus* (fig. 76) and *Agriornis microptera* (fig. 78). Thus *D. g. domeykoanus* can be reckoned as a faunal element of the Chilean subcentre of the Andean Pacific centre, *D. g. gracilis* as a Monte faunal element, *D. g. griseus* as a Patagonian faunal element and *D. g. maullinicus* as a *Nothofagus* faunal element. *D. g. maullinicus* is the only faunal element of the *Dusicyon* group that is adapted to forest and

Fig. 77. Vultur gryphus and *Cinclodes fuscus* as examples of certain species which are mainly distributed in the Andes above 3000 m. but which come down to sea level in the Patagonian centre.

therefore constitutes an exception. The others only occur in open landscapes.

The Andes are a very clear line of separation for most of the vertebrate groups dealt with here. It is only possible to speak of an undivided Patagonian sub-region when considering taxa which on the one hand are of pre-Andean age and evolving at a very slow rate, and on the other are not adapted to any particular vegetational formation (cf. EIGENMANN 1905, GERY 1969, NOODT 1969). The dispersal centres of such taxa will only be congruent with those of the groups dealt with here when the position of the dispersal centres described in this book is controlled geologically and not by fluctuations in vegetation e.g. Galapagos centre, Talamanca centre, cf. Discussion.

The Pre-Pliocene fauna of the Patagonian centre is the best known in all South America (SCHAEFFER 1949, CASAMIQUELA 1958, ROMER 1966). It agrees with that of the Monte centre and can be interpreted in the same way. In contrast to the Pampa centre, the Pliocene and Pleistocene faunas are less well known.

Among Argentinian bird species at least 36 should be reckoned as endemic to the centre i.e. as stationary faunal elements in the sense of DE LATTIN. This statement is based on the work of OLROG (1963, p. 31).

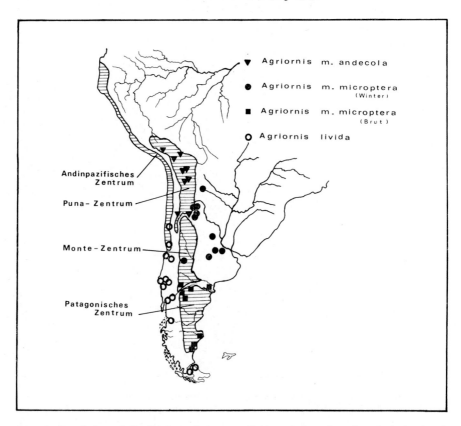

Fig. 78. Correlation of distribution of *Agriornis livida* and the subspecies of *Agriornis microptera* with the Patagonian, Monte and Puna centres (cf. VUILLEUMIER 1972).

40. The Nothofagus centre

The position of the centre (fig. 5, no. 40) can be defined on the ranges of the monotypic amphibian genera *Telmatobufo*, *Batrachyla*, *Hylorina*, *Calyptocephalella* and *Rhinoderma* (fig. 79), and of the monotypic bird genera *Sylvior-*

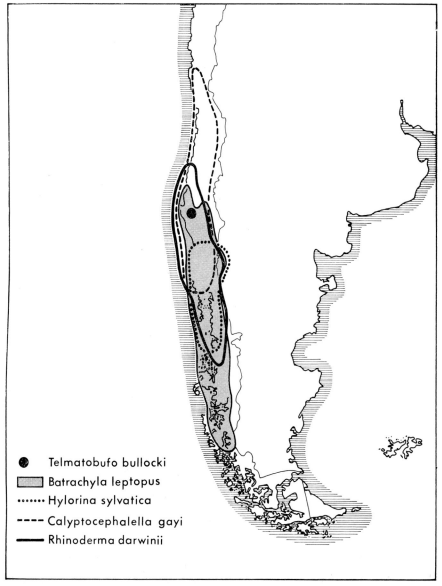

● Telmatobufo bullocki
▨ Batrachyla leptopus
······ Hylorina sylvatica
---- Calyptocephalella gayi
—— Rhinoderma darwinii

Fig. 79. Ranges of five faunal elements of the *Nothofagus* centre, all monotypic amphibian genera. Locality data from CEI (1962).

155

thorhynchus, Aphrastura, Pygarrhichas and *Enicognathus* (fig. 80). Moreover all Chilean species of *Eupsophus* (fig. 81) can be considered as faunal elements of the centre (for the definition of the genus *Eupsophus* cf. CEI & ERSPAMER 1966, MÜLLER 1970).

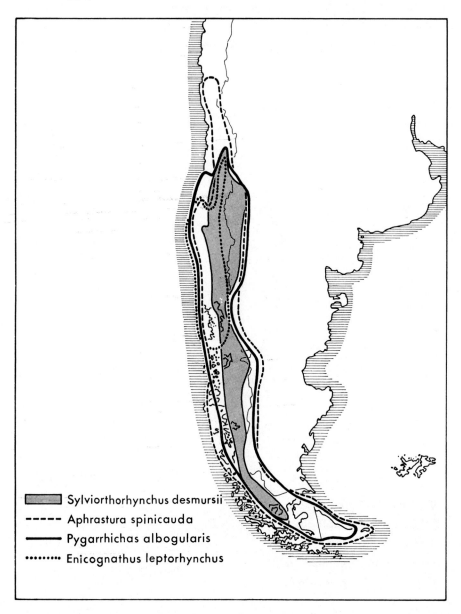

Sylviorthorhynchus desmursii
----- Aphrastura spinicauda
——— Pygarrhichas albogularis
········· Enicognathus leptorhynchus

Fig. 80. Distribution of four faunal elements of the *Nothofagus* centre — all monotypic bird genera.

156

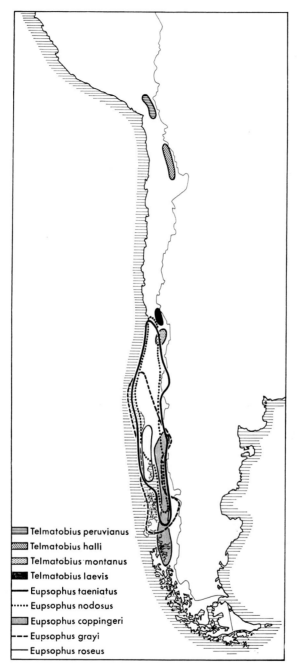

Telmatobius peruvianus
Telmatobius halli
Telmatobius montanus
Telmatobius laevis
Eupsophus taeniatus
Eupsophus nodosus
Eupsophus coppingeri
Eupsophus grayi
Eupsophus roseus

Fig. 81. Chilean *Eupsophus* species can all be considered as faunal elements of the *Nothofagus* centre. On the other hand Chilean *Telmatobius* species are adapted to high mountains and can be derived from species of the Puna centre.

The faunal elements of the centre are the essential cause of the faunistic distinctiveness of Chile. This distinctiveness was already recognised long ago (GOETSCH 1930, 1933, GOETSCH & HELLMICH 1932, 1933, HELLMICH 1933, 1934, 1938, MÜLLER & HELLMICH 1932, MÜLLER 1938). But more recently, authors who have sought to erect regions defined by contained genera or families have felt justified in uniting the Chilean and Argentinan areas together (CABRERA & YEPES 1940, MELLO-LEITÃO 1946, HERSCHKOVITZ 1958, 1969, FITTKAU 1969, GERY 1969, KUSCHEL 1969). This unification of the two areas into a single Patagonian subregion basically depends on two things. Firstly it is a statistical expression of the richness in species of relatively old invertebrate groups, which in many cases have no special preference for a particular type of forest. Secondly it depends on the presence of a large number of young groups with strong powers of movement, such as birds, which have a correspondingly wide ecological valency. For a qualitative study, such as the analysis of dispersal centres, these types of groups are not very important. For other zoogeographical problems, of course, such as proving land-bridges across the South Pacific, their value is undeniable (cf. HENNIG 1960, ILLIES 1960, 1961, 1965, 1966, 1969, KUSCHEL 1960, NOODT 1965, 1967, 1969, BRUNDIN 1966, 1972, MÜLLER & SCHMIT-HÜSEN 1970, THENIUS 1969, 1972).

This does not mean that every group of animals has its own zoogeography. It only means that the ecological valency of individual species of particular families has been acquired in different biotopes and that the biotopes in question correspond to the particular ecological valency. Consequently the distribution of each species must be considered in connection with the evolution of the area that it lives in. For the distribution of fishes, as I showed when discussing the Costa Rica centre, quite other factors are critical than for the birds, in accordance with the ecological valency of fishes (cf. GERY 1969). It is therefore entirely mistaken and arbitrary to use both fishes and birds in defining a region.

For phylogenetically old taxa, such as fishes (GERY 1969), with a relatively slow rate of evolution and living in water, or subterraneously, it may possibly be justified to unite Chile and Patagonian Argentina into a single centre. After careful consideration of the ecological valency of the animals that show it, the existence of this type of distribution covering both Chile and Patagonian Argentina could be taken as indicating a single, original, Chilean-Argentinian faunal area that existed before the rise of the Andes. But present-day terrestrial vertebrate species are adapted either to a forest or to a steppe biotope and none of them has this type of distribution.

Among invertebrates an extremely ancient type can be recognised whose phylogeny was probably connected with the development of the southern Pacific *Nothofagus* forests. Thus the distribution of the Carabid tribe of Migadopini (DARLINGTON 1960) is essentially like that of the genus *Nothofagus* (SCHMITHÜSEN 1953, 1966, 1968). And its evolution may perhaps have been correlated with the evolution and dispersal of *Nothofagus* as discussed mainly by geographers and botanists (e.g. DUSEN 1899, FLORIN 1940, DAWSON 1958, COUPER 1960, GODLEY 1960, HOLDGATE 1960, KNOX 1960, SCHMITHÜSEN 1966).

To clear up this question properly, however, will require further work. The essential features of the herpetofauna of Chile are known through the work of: GUICHENOT (1848), GIRARD (1854), LATASTE (1891), WERNER (1896, 1897), ANDERSON (1898), PORTER (1898), PHILIPPI (1902), ROUX (1910), QUIJADA (1911, 1914), BARROS (1918), KRIEG (1924), SCHNEIDER (1930), HELLMICH (1933), NOBLE (1938), SCHMIDT (1952, 1954), CAPURRO (1953, 1954, 1957, 1960), CEI (1957, 1958, 1959, 1960, 1961, 1962), DONOSO-BARROS (1961), BARROS & CEI (1962) and GORMAN (1968). Real surprises, which would affect the zoogeographical picture decisively, are no longer to be expected. Only 19 species of amphibia occur in the Chilean area, divided between nine genera. Of these, 13 species and one subspecies should be reckoned as *Nothofagus* faunal elements. Of the nine genera, only five are monotypic endemic genera of the *Nothofagus* centre, i.e.: *Telmatobufo bullocki, Batrachyla leptopus, Hylorina sylvatica, Calyptocephalella gayi* and *Rhinoderma darwini*.

A complete list of Chilean amphibia, with *Nothofagus* faunal elements marked (N), reads as follows:

Bufonidae
Bufo spinulosus spinulosus (cf. fig. 82)
B. spinulosus atacamensis
B. spinulosus arunco
B. spinulosus rubropunctatus (N)
B. variegatus (N)
Leptodactylidae
Telmatobufo bullocki (N, cf. fig. 79)
Telmatobius peruvianus (cf. fig. 81)
T. halli
T. montanus
T. laevis
Eupsophus taeniatus (N, cf. fig. 81)

E. nodosus (N)
E. coppingeri (N)
E. grayi (N)
E. roseus (N)
Batrachyla leptopus (N, cf. fig. 79)
Hylorina sylvatica (N, cf. fig. 79)
Pleurodema bibroni (cf. fig. 82)
P. bufonina (N)
P. marmorata
Calyptocephalella gayi (N, cf. fig. 79)
Dendrobatidae
Rhinoderma darwinii (N, cf. fig. 79)

The *Nothofagus* centre therefore has a high proportion of endemic genera. The origin of the monotypic genera in no case implies connections with New Zealand or Australia. The indigenous New Zealand amphibian fauna is poor in species and completely isolated in systematic position. According to GORHAM (1966) it includes only *Leiopelma archeyi, L. hochstetteri* and *L. hamiltoni*. (In passing, it seems extremely doubtful that there are monophyletic connections of the New Zealand fauna with *Ascaphus truei* of western North America.)

The complete absence of the family Hylidae in Chile is striking. The Dendrobatidae are a purely Neotropical group. The closest relatives of *Rhinoderma* occur in the Brazilian rain-forest biome (Amazon centre). The Leptodactylidae are represented at the present day in both Australia and South America. The Chilean species show no close connection with those of Australia.

Some of the amphibia that occur in the *Nothofagus* centre, have direct connections with forms widespread in northern South America, belonging to the same or closely related species, or closely related genera; this group comprises

159

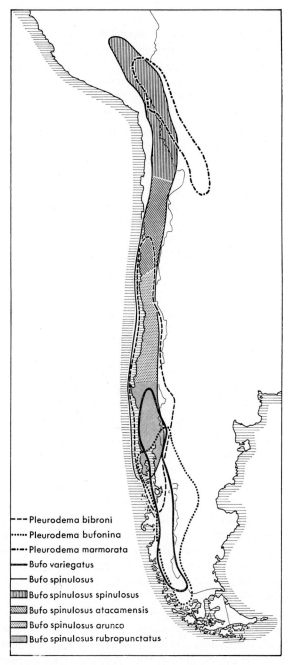

Fig. 82. Distribution of Chilean *Pleurodema* and *Bufo* species. Only *P. bibroni*, *B. variegatus* and *B. spinulosus rubropunctatus* can be regarded as *Nothofagus* faunal elements. *P. marmorata* and *B. spinulosus spinulosus* are Puna faunal elements. And *B. spinulosus arunco* must be ascribed to the Chilean subcentre of the Andean Pacific centre. *Pleurodema bufonina* exists also in eastern *Patagonia*.

160

Telmatobius, *Eupsophus* and *Pleurodema*. Others arose from a basic stock that diverged from recent Neotropical genera; this group includes *Telmatobufo*, *Batrachyla*, *Hylorina* and *Calyptocephalella*.

Pre-glacial fossil history shows that forms closely related to the above-mentioned Leptodactylid genera were already present in the Tertiary of Patagonia. Such forms include *Eophractus* from the Eocene, *Gigantobatrachus* from the Miocene and *Calyptocephalella canqueli* from the Oligocene (cf. VUILLEUMIER 1968). It follows that not all *Nothofagus* faunal elements need have immigrated from the north into the *Nothofagus* region, although CEI (1962) and VELLARD (1957) have assumed so.

BLAIR (1963) supposed that the Neotropical Bufonids were relatively new immigrants from the Nearctic, by way of the Central American land bridge. This idea, however, also needs revision (TIHEN 1962, MÜLLER 1968). ESTES & WASSERSUG (1963) found a form in the Miocene beds of the Magdalena graben of Colombia which was amazingly similar to the present-day *Bufo marinus*. Furthermore a Bidder's organ has been discovered in a whole series of anurans formerly placed in other Neotropical families as for example in *Dendrophryniscus brevipollicatus* (cf. MÜLLER 1968). Forms having this organ must now be placed in the Bufonids. The view of VELLARD (1957) and CEI (1962), that the two Chilean Bufonid species were both new immigrants, must be partly false purely on distributional grounds.

Thus *Bufo variegatus* has certainly been limited to the *Nothofagus* forests since at least the beginning of the Pleistocene. Only *Bufo spinulosus* must be considered a recent northern immigrant (fig. 82), probably derived from the Nearctic *B. valliceps* group. It has reached the *Nothofagus* region by following the Andean chain southwards. This species can be divided into four subspecies and shows a shift in vertical range according to climatic conditions. In the Peruvian Andes of the Puna centre *B. spinulosus* is still met with above 4000 m. but in the southern part of its range, as the subspecies *B. s. rubropunctatus*, it almost comes down to sea level, as for instance in the island of Chiloe. MANN (1968, p. 185) reckoned *B. spinulosus* as a member of the 'hochandine Lebensgemeinschaft', but this is basically true only for the northernmost subspecies.

Amphibian faunal elements of the *Nothofagus* centre thus all show very similar distributions at the present day, but nevertheless began to differentiate at very different times. This appears not only from a study of the recent distribution but also from palaeontology and from evolutionary genetics. It is also possible to show that the *Nothofagus* centre has been the centre of origin for some taxa. Among the monotypic genera of the *Nothofagus* forests, *Calyptocephalella* has been recorded in the Oligocene of Patagonia (*Calyptocephalella canqueli*).

A revision of the supposed *Eupsophus* species occurring in the Serra do Mar centre shows that they belong to a separate genus from the *Nothofagus* populations and should be separated again under the name *Thoropa* (MÜLLER 1970). The electrophoretic results of CEI, ERSPAMER and ROSEGHINI (1967) support this conclusion.

The results got from the amphibia can be confirmed from the reptiles. Unfortunately the species of *Liolaemus*, which in general are very suitable for zoogeographic analysis (cf. HELLMICH 1934), have not been sufficiently studied from the systematic viewpoint (see also DONOSO-BARROS 1961, 1970). Also, with a few exceptions (fig. 83, 84), they are not specially well suited for a zoogeographic

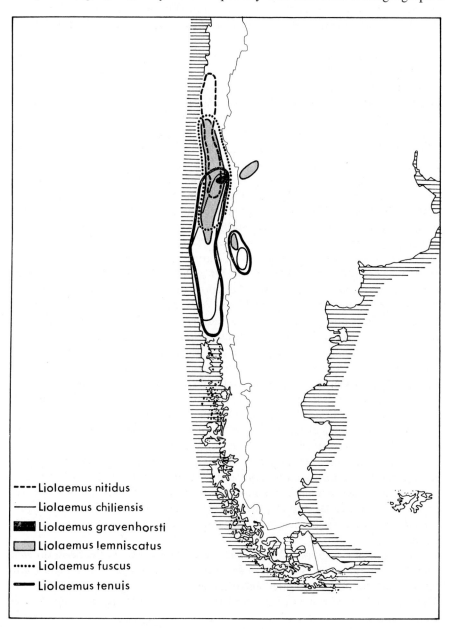

```
---- Liolaemus nitidus
──── Liolaemus chiliensis
▓▓ Liolaemus gravenhorsti
▒▒ Liolaemus lemniscatus
•••••• Liolaemus fuscus
▬▬ Liolaemus tenuis
```

Fig. 83. Distribution of six Chilean *Liolaemus* species (Iguanidae). Only two of these (*L. tenuis* and *L. chiliensis*) can be regarded as *Nothofagus* faunal elements.

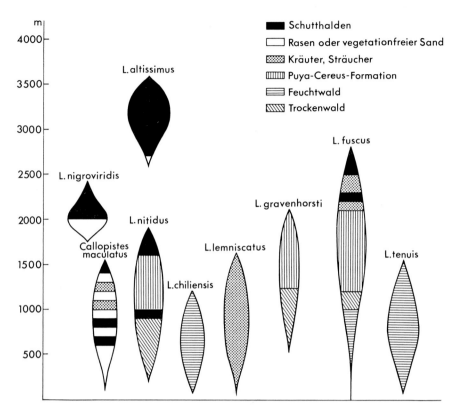

m

4000

3500

3000

2500

2000

1500

1000

500

L. altissimus

L. nigroviridis

L. nitidus

Callopistes
maculatus

L. chiliensis

L. lemniscatus

L. gravenhorsti

L. fuscus

L. tenuis

■ Schutthalden
☐ Rasen oder vegetationfreier Sand
▨ Kräuter, Sträucher
▥ Puya-Cereus-Formation
▤ Feuchtwald
▧ Trockenwald

Fig. 84. Vertical distribution and preferred biotope of eight *Liolaemus* species (Iguanidae) and of *Callopistes maculatus* (Teiidae) in the region round Santiago de Chile (data from HELLMICH 1933).

analysis of the forest fauna. On the other hand the Iguanids *Urostrophus valeriae* and *Urostrophus torquatus* (fig. 85) are typical faunal elements of the *Nothofagus* centre. In accordance with their preference for high temperatures reptiles become less and less important, at least within the *Nothofagus* centre, going south towards the Magellan straits.

The mountain passes of the southern part of the *Nothofagus* region in Chile and Argentina are relatively low-lying and do not rise above the tree-line. For this reason some *Nothofagus* faunal elements that range southwards, also extend eastwards through the passes on to Argentinian soil. As HELLMICH remarked (1933) the 'gaps' in the Patagonian Cordillera between 38° and 45°S. have had a similar effect. Whether those gaps still serve for faunal exchange (e.g. north of Nahuel Huapi), is still not clear. But it does appear that the populations inhabiting the *Nothofagus* biome east and west of the Cordillera, are only subspecifically distinct from each other, if at all. Pollen-analytical results confirm that immigration took place only a short time ago i.e. in postglacial times (AUER 1960, p. 513).

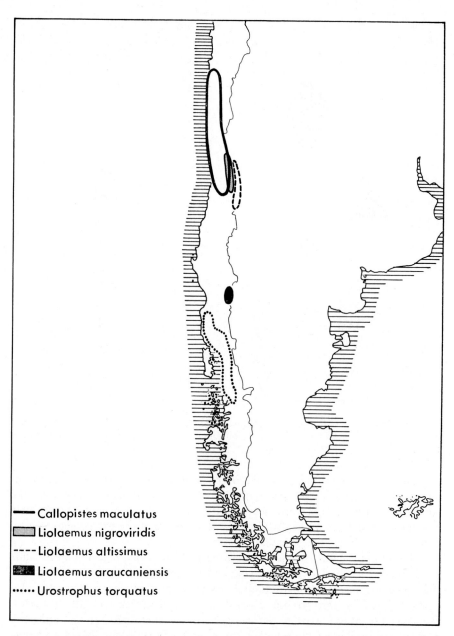

Callopistes maculatus
Liolaemus nigroviridis
Liolaemus altissimus
Liolaemus araucaniensis
Urostrophus torquatus

Fig. 85. Callopistes maculatus is restricted in range to the Chilean subcentre of the Andean Pacific centre. The Iguanid *Urostrophus torquatus* on the other hand is strictly limited by ecology to the forests of the *Nothofagus* centre. The distribution of the *Liolaemus* species *L. nigroviridis*, *L. altissimus* and *L. araucariensis* shows that the species of the genus can be used as extremely sensitive ecological indicators.

Passing from the herpetofauna to the birds, it appears that at least 37 of the populations occurring in the *Nothofagus* forests are endemic, either as species or subspecies. This is based on data in GOODALL, JOHNSON & PHILIPPI (1946–1964), PHILIPPI (1964), JOHNSON (1965), VUILLEUMIER (1967). According to MEYER DE SCHAUENSEE (1966), five of these endemics must be considered as monotypic genera. Birds are very poorly known as fossils and because of their powers of flight must be treated very cautiously in zoogeographical analysis, so it is not possible to be so definite as with the amphibia. Nonetheless it seems very likely that the three monotypic Furnariid genera *Sylviorthorhynchus*, *Aphrastura* and *Pygarrhichas* have always been linked in phylogeny to the *Nothofagus* forests. Of course, as shown in fig. 80, *Aphrastura spinicauda* does occur in places outside the *Nothofagus* region.

These *Nothofagus* faunal elements in the avifauna all show obvious morphological divergences from the taxa closest related to them. It can only be supposed, therefore, that for the birds also the *Nothofagus* centre has long been isolated from other forest areas.

The mammal fauna of the *Nothofagus* region fits into the same picture without any contradictions, although unfortunately it is in no way so well studied as the bird fauna (OSGOOD 1943, CABRERA 1957, etc.). Two monotypic marsupial genera of the *Nothofagus* forests i.e. *Rhyncholestes* and *Dromiciops* have a long fossil history. Thus *Rhyncholestes* can be followed back into the Eocene of Patagonia while *Dromiciops* occurs in the Oligocene—Miocene of Patagonia along with related genera. *Pudu pudu*, *Akodon sanborni* and the monotypic rodent genus *Irenomys* are confined to the Nothofagus centre.

As GOETSCH (1933) and others showed for the terrestrial planarians, the same relationships as for the vertebrates can be established in many intervertebrate groups.

The fact that the nucleus of the dispersal centre is displaced to the north relative to the total extent of the South American *Nothofagus* forests, can partly be understood from present-day ecology (HETTNER 1881, HAUMAN 1913, KNOCHE 1927, BRÜGGEN 1934, LLIBOUTRY 1956, SCHWABE 1956, 1969, DAHL 1960). From the regional point of view the nucleus includes the 'Valdivian forest' region of KUSCHEL (1960). But as already explained, the faunal elements of the centre extend beyond this nucleus towards the south.

It is certain that the position of the dispersal centre partly results from climatic oscillations during the Pleistocene and post-glacial and from vegetational fluctuations correlated with them. This follows from the work of CUNNINGHAM (1871), BRÜGGEN (1929, 1948), SAHLSTEIN (1932), BERRY (1937), SALMI (1941), AUER (1960) and SCHMITHÜSEN (1966). The position of the *Nothofagus* centre nonetheless indicates that the Valdivian rain-forest region has not been affected by such vegetational movements.

KUSCHEL (1960) has drawn attention to connections between the 'Valdivian and southern Brazilian forest'.

The affinity between two or more dispersal centres depends on the phylogeny of the faunal elements ascribed to them (MÜLLER 1971, 1972). The phylogenetic relations of the faunal elements of the Neotropical dispersal centres have already been discussed in the previous section of the book, in so far as the evolution of the individual taxa has been worked out.

I shall now use the results of the previous section to arrange the dispersal centres into **affinity groups.** I shall pay special attention to polytypic and polycentric species whose ecological valency is known. On the other hand I shall not consider supraspecific units i.e. the ranges of genera or families. Such groups are often phylogenetically very old (cf. MAYR 1967, DE LATTIN 1967) and the individual taxa within them have had time to adapt to very different biotopes. At the species level, and usually at the subspecies level, the type of distribution can be correlated with the particular vegetational formations or climatic types to which the elements are adapted.

I shall now introduce ten different examples and explain them more closely.

1. *Lachesis mutus*—the bush master—is a polytypic pit viper (Crotalidae), which is ecologically strictly adapted to the lowland rain forests of Central America and northern South America (cf. HOGE 1966, MÜLLER 1968). The species has never yet been seen outside tropical rain forests. Its monocentric subspecies are faunal elements of the Serra do Mar centre (*L. m. noctivagus*), of the Amazon centre (*L. m. mutus*) and of the Costa Rican centre (*L. m. stenophrys*, cf. fig. 70).

2. *Crotalus durissus* is a polytypic species of rattlesnake which, in contrast to *Lachesis mutus*, strictly avoids rain forest (HOGE 1966, MULLER 1970). If the ranges of the subspecies of this Crotalid are considered (fig. 86), two are found on islands i.e. *C. d. marajoensis* on Marajó and *C. d. unicolor* on Aruba; one subspecies (*C. d. dryinus*) is endemic to the coastal savanna of Guyana; and the rest are faunal elements of dispersal centres 3, 4, 8, 19, 20, 31, 32 and 34. Subspecifically distinct populations have to be allopatric in distribution for reasons of evolutionary genetics. Consequently the respective dispersal centres are also centres of differentiation for the populations in question.

3. *Campylorhynchus griseus* is a polytypic bird species of the family Troglodytidae. Like *Crotalus durissus* it avoids the rain forest (fig. 87). Its monocentric subspecies are faunal elements of centres 4, 8, 11, 19 and 20. A subspecifically distinct population also occurs in the campo 'islands' west of Monte Duida in Venezuela.

4. *Sanguinus midas* is a polytypic species of monkey with obvious adaptational mechanisms for life in rain forest. Its two monocentric subspecies are faunal elements of centres 22 and 23. The distribution of the other species of *Sanguinus* (fig. 88), except for *S. oedipus*, is correlated with centre 25 (Amazon centre). The polytypic species *S. oedipus* in the nominate form is a faunal element of centre 8, together with *Crotalus durissus* and *Campylorhynchus griseus*. The subspecies *S. oedipus geoffroyi* is a faunal element of the Choco subcentre of centre 14.

Fig. 86. Distribution and phylogenetic relationships of the subspecies of the South American rattlesnake *Crotalus durissus*. The numbers inside the circles refer to the numeration of the dispersal centres in fig. 5, and in the text. Thus the subspecies *C. d. collilineatus* is a faunal element of the Campo Cerrado centre, etc.

In discussing the faunal elements of centre 8 (Barranquilla centre) I have already pointed out that most of its faunal elements are closer related to those of centres 4, 5, 12, 11 and 19 than to centres 9 or 14. *Sanguinas oedipus* shows, therefore, that species with suitable preadaptations are not always bound by the generally valid scheme of relationships between centres.

5. The polytypic Fringillid *Atlapetes brunneinucha* is represented by mono-centric subspecies in the centres 2, 13, 18 and 26 (fig. 89). Occurences of breeding birds outside montane forest are not at present known. The ranges of the sub-

Fig. 87. Ranges of the subspecies of the polycentric and polytypic Troglodytid species *Campylorhynchus griseus*. The monocentric subspecies of this species are faunal elements of the centres and a subspecifically distinct population occurs on the campo islands west of Monte Duida, Venezuela.

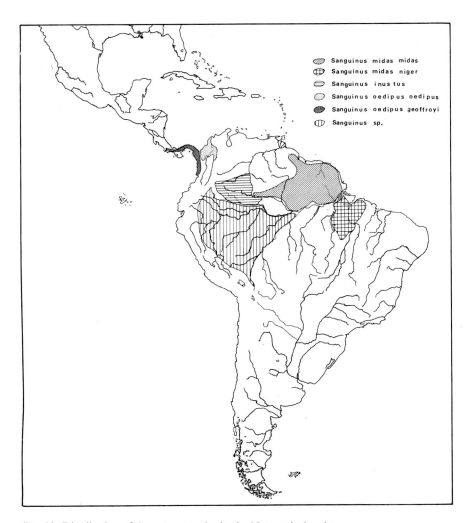

Fig. 88. Distribution of *Sanguinus* species in the Neotropical realm.

species therefore indicate a direct relationship between centres 2, 13, 18 and 26.
6. The Tinamid *Rhynchotus rufescens* is a polytypic and polycentric species.
Its subspecies *R. r. catingae* is a faunal element of the Caatinga centre, *R. r.
rufescens* of the Campo Cerrado centre, and *R. r. pallescens* of the Pampa
centre (fig. 90). In the westernmost part of the Chaco region the subspecies
R. r. maculicollis occurs, and the populations living in the campo of Marajó
can be derived from *R. r. rufescens*.
7. *Crax rubra* is a polycentric group of birds strictly adapted to forest. It forms
a super-species complex of seven semispecies (VUILLEUMIER 1965) (fig. 91). These

Fig. 89. The range of the polytypic Fringillid *Atlapetes brunneinucha* — a species of the montane
forest.

Fig. 90. Distribution of *Rhynchotus rufescens* — a polytypic Tinamid species adapted to open landscape.

seven semispecies are monocentric for the dispersal centres 14, 17, 22, 24, 25 and 33.

8. The monkey genus *Callithrix* (fig. 92) is represented by two endemic species in the Madeira centre and five in the Serra do Mar centre. However *Callithrix penicillata penicillata*, which occurs in the Serra do Mar centre, is closest related to a population in the Campo Cerrado centre (*C. p. jordani*). This last subspecies in places occurs in the Campo Cerrado sympatrically with *Callithrix argentata melanura*. And this species (*C. argentata*) penetrates to the south bank of the Amazon having two subspecies—*C. a. leucipe* and *C. a. argentata*—in the isolated 'campo islands' between the rivers Tapajoz and Tocantins.

Fig. 91. Ranges and phylogenetic connections (arrows) of the semispecies of the *Crax rubra* superspecies complex (data from VUILLEUMIER, 1965).

9. The polytypic hare species *Sylvilagus floridanus* has a distribution like that of *Campylorhynchus griseus*. It occurs in the Caribbean centre as the subspecies *S. f. continentis*, in the Magdalena centre as the subspecies *S. f. purgatus* and in the Barranquilla centre as the subspecies *S. f. superciliaris*.

10. *Sciurus aestuans* is a polytypic species of squirrel adapted to the forest biome. Its subspecies are faunal elements of the Serra do Mar centre (*S. a. alphonsei, S. a. garbei, S. a. ingrami, S. a. henseli*); of the Para centre (*S. a. paraensis*); of the Guyanan centre (*S. a. aestuans*); of the Pantepui centre (*S. a. quelchii, S. a. macconelli, S. a. venustus*); and of the Amazon centre (*S. a. gilvigularis*).

C. humeralifer	⊜
C. chrysoleuca	---
C. aurita	⊘
C. geoffroyi	----
C. flaviceps	....
C. jacchus	⊞
C. p. penicillata	⊘
C. p. jordani	⊞
C. a. argentata	•
C. a. leucipe	○
C. a. melanura	□

Fig. 92. Distribution of the species of the monkey genus *Callithrix*. Fuller explanation in text.

The centres can be arranged in **affinity groups** according to the degree of relationship of their faunal elements, especially at the subspecific level. These groups are:
Group I: 3, 4, 5, 8, 11, 12, 19, 20, 28, 29, 30, 31, 32, 34, 35, 36, 37, 38 and 39.
Group II: 1, 2, 6, 9, 10, 13, 14, 16, 17, 18, 21, 22, 23, 24, 25, 26, 33 and 40.
Group III: 7, 15 and 27.
The distinction between groups I and II is very obvious in most cases. The only exceptions are centre 8 (Barranquilla centre) and centre 32 (Campo Cerrado centre). In these two centres a sort of interpenetration of group I and II occurs, since faunal elements of both affinity groups can be ascribed to the centres (cf. fig. 88, 92). Both 8 and 32, however, should certainly be placed in group I

173

Fig. 93. The arboreal dispersal centres of terrestrial vertebrates in the Neotropical realm. The montane forest centres are shown by fine hatching and the lowland forest centres by coarse hatching (cf. fig. 3, 88, 89, 91). Fuller explanation in text.

because they have more than 60% of their elements directly related to faunal elements of group I centres (cf. discussion of the Campo Cerrado and Barranquilla centres).

Comparable transitional relationships never exist between affinity groups II and III. As against this there is a tight interpenetration of groups I and III in centres 29, 37 and 39 (i.e. Andean Pacific, Monte and Patagonian centres). There is a particularly close relationship between centre 27 in group III (Puna centre) and centre 39 in group I (Patagonian centre).

Faunal elements of the Patagonian centre in most cases are closely related to

Fig. 94. The non-forest dispersal centres of terrestrial vertebrates in the Neotropical realm (cf. fig. 3, 87, 90). Fuller explanation in text.

those of the Monte centre (37) and Pampa centre (38), and these in turn are closely related to the elements of the Uruguay (35) and Chaco centres (36). This is the reason for placing the Patagonian centre in group I.

If the three affinity groups are plotted separately from each other on a map of the Neotropical realm, three things become obvious (cf. fig. 93, 94, 95 and 96). These are:

1. The centres belonging to group I are connected with unforested areas (fig. 94), or with at least partly unforested areas below 1500 m. i.e. campos cerrados, savannas, deserts etc.

2. The centres belonging to group II lie in rain-forest areas (fig. 93) to which their faunal elements are adapted.

Fig. 95. The oreal distribution centres of terrestrial vertebrates in the Neotropical realm (cf. fig. 3). Fuller explanation in text.

3. The centres belonging to group III lie above the tree-line, in the Andes (fig. 59).

It appears that the distribution of the faunal elements of groups I, II and III is correlated with three major environments. But each of these major environments only represents an outer frame that includes a great amount of differentiation within itself. There are some difficulties in equating the concept of these major environments with the biochore concept that DE LATTIN applied to the Holarctic region (1967, p. 246).

SCHMITHÜSEN (1968, p. 235) understands biochores as being: 'environmental units of the landscape'.* DE LATTIN (1967, p. 246), on the other hand, said that:

* 'lebensräumliche Einheiten der Landschaft.'

Fig. 96. Distribution of rain forests (cross-hatched) and open landscapes (blank) in the Neo-tropical realm (after SCHMITHÜSEN 1968). Note the isolated campo islands in the Amazonian forest area.

'Zoogeographical subdivision into biochores appears by contrast to be prima facie simpler, not to say coarser [than phytogeographical subdivision into biomes]. This is no doubt connected with the powers of locomotion of animals, with their greater independence of many environmental factors important for plants, and probably also with their greater adaptability to ecologically varying conditions. As a result a gross division of the world of land animals leads to the separation of only three major environments: the arboreal, the eremial and the oreotundral.'*

* 'Die zoogeographische Gliederung in Biochoren erweist sich demgegenüber als einfacher, um

DE LATTIN considered, however, that this gross subdivision only held for non-tropical areas (1967, p. 80, 81). In tropical areas, on the other hand, he separates arboreal, savanna and eremial biochores from each other. CHAPMAN (1926) and HAFFER (1967, 1968) include savannas and deserts together as 'non-forest' and oppose them to the arboreal or rain-forest environment which they call 'forest'.

Basically, however, the concepts mentioned in the last paragraph are un-satisfactory. In what follows I shall therefore hold mainly to the terminology that de Lattin applied only to non-tropical areas. I shall denote the animals of group I as the fauna of the **non-forest** environment (biome; see CARPENTER 1939), those of group II as the fauna of the **arboreal** environment and those of group III as the fauna of the **oreal** environment.

The **arboreal** environment as I conceive it, however, only includes the rain-forest biome, comprising both lowland and montane rain forest. It does not include arid forest, campos cerrados or humid savanna. The **arboreal** concept is there-fore used in a narrower sense than the word arboreal suggests. This necessarily leads to a widened definition of the non-forest biome. I will nonetheless retain the concept of an arboreal biome on methodological grounds.

In DE LATTIN's sense the eremial biochore equals arid steppe plus desert. As I conceive it, however, the non-forest environment it includes not only the deserts and Patagonian cold steppe but also the arid forests, campos cerrados and humid savanna. Placing these latter vegetational formations in a non-forest environment is justified because of the high degree of relationship in their faunas. Compare in this connection the discussion of the Andean Pacific and Chaco centres, etc.

The oreal environment is conceived here exactly as DE LATTIN defined it. It in-cludes the environments above the montane forest (cf. ELLENBERG 1966).

The potential range of action of the faunas of the three major environments is constrained by the strict ecological adaptation of the animals. For example, non-forest landscapes lie outside the range of action of flying, rain-forest birds; rain-forest areas of high precipitation are outside the range of action of campo species such as *Crotalus durissus*; and regions of cold climate at high altitudes are completely outside the range of action of tropical lowland species.

Natural barriers of these sorts are mainly responsible for the discontinuities between geographical isolates and decide, or at least influence, the dispersal rate of a taxon. This dispersal rate is a measure of the effectiveness of the iso-lation of a population. Usually this can only be deduced indirectly from the level of differentiation of populations isolated in space, apart from a very few exceptions (MACARTHUR & WILSON 1967, WILSON & SIMBERLOFF 1969).

I have pointed out elsewhere (MÜLLER 1966), that separating barriers of water have a much stronger isolating effect for forest birds than for non-forest

nicht zu sagen gröber, was zweifellos mit der Lokomotionsfähigkeit der Tiere, mit deren stärkeren Unabhängigkeit von manchen für die Pflanzen wichtigen Umweltfaktoren und wohl auch mit deren schnellerer Anpassung an ökologisch wechselnde Bedingungen zusammenhängt. Demgemäß führt eine Großgliederung der Landtierwelt in Biochoren eigentlich nur zur Unterscheidung von drei Großlebensräumen: dem Arboreal, dem Eremial und dem Oreotundral.'

species (see also SNETHLAGE 1910, 1913). This result is confirmed by the faunal elements of the arboreal centres 22, 23, 24 and 25 within Amazonia. Also two subcentres i.e. the Napo and Ucayali subcentres, can be distinguished within the Amazon centre itself on the basis of the isolating effect of the Amazon-Solimões river. These two subcentres are nonetheless much better marked among birds and mammals than in amphibia or reptiles. In the same way all the other big rivers in the Amazon area are important as isolating factors for the forest fauna. This is true for the Rio Madeira and Rio Negro among others, as shown by the ranges of *Penelope marail*, *Psophia viridis* and *Mitu mitu tuberosa*.

There is an obvious barrier-line for the faunal elements of centres 22, 23, 24 and 25 approximately at the 1500 m. contour. Lowland species only cross it in regions specially favoured by edaphic factors (cf. Central American rain-forest centre). But the 1500 m. contour is not only important for lowland species. It is equally effective for montane forest species, though among birds mostly only during the breeding season. Rivers are totally unimportant as barrier-zones for the faunal elements of montane forest centres. This is obviously due to the relative narrowness of rivers in their upper reaches. Both lowland forest and montane forest species, however, have in common an additional limit to distribution, which has not yet been given enough attention in South America. This limit is set by the 'open landscapes' i.e. campos, campos cerrados, savannas, puna and paramo areas and deserts.

Obviously a whole series of other isolating factors exist, including 'psychological' ones (cf. MAYR 1967). But for faunas of arboreal, neotropical distribution centres the most important isolating barriers are river systems, arms of the sea, the 1500 m. contour and open landscapes. It may be that placing the altitude-barrier at 1500 m. is over-simple, but I have come to this particular value in the tropical area because the available localities for lowland forest and montane forest faunal elements reveal a barrier-line at this height.

Non-forest species, conversely, have an obvious distributional limit where they meet the rain-forest biomes. Faunal elements of the non-forest centres do not penetrate the arboreal centres unless the latter have been made treeless by man cf. MÜLLER (1970). Rivers and seas still have an isolating effect in the non-forest biochore, but much less than with forest birds (fig. 97). Also the 1500 m. contour clearly does not form such an obvious limit in the distributional picture as it does with the forest fauna. Compare in this connection the similar faunas of the Patia and Cauca grabens.

The oreal species are almost confined to the treeless environment of the Andean mountain system and penetrate only rarely into the highest altitudinal zone of the cloud forest.

Despite all that I have said, the limits of a whole series of arboreal centres of the neotropical region cannot be explained by the barriers that usually isolate the forest faunas, at least in the form that those barriers now have. This is true, for instance, of the boundaries of centres 1, 6, 14, 23 and 24. Again, the present-day ecology in no way explains the complete lack of an arboreal dispersal

179

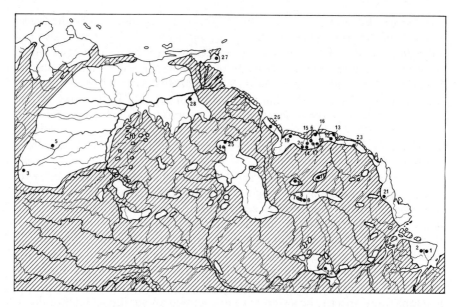

Fig. 97. Distribution of the savanna species *Kentropyx striatus* (rain forest = cross hatched).

centre between the rivers Xingu and Tocantins, the different degrees of differentiation of the Nechi and Choco faunal elements or the limit between the Central
American rain-forest centre and the Costa Rican centre.

A further problem is the different degree of differentiation of the faunal
elements of the individual centres. Thus the populations of the Santa Marta
centre, with one exception, are only subspecifically distinct. As against this, the
Pantepui and *Nothofagus* centres have a high percentage of endemic species
and monotypic genera.

An attempt to subdivide the three affinity groups defined above shows that
the degree of relationship between centres does not only depend on the ecological distinctiveness of a centre within a group. It equally depends on the
vagility, the course of evolution and the geographical isolation of the faunal
elements of the centres.

Within the first or non-forest group the following centres show the greatest
affinity with each other.

    a. 3, 4, 5.
    b. 8, 11, 12, 19, 20.
    c. 31, 32, 34, 36.
    d. 35, 37, 38, 39.
    e. 28, 29, 30.

Likewise within the second or arboreal group the following centres show
greatest affinity.

    a. 1, 6, 9, 14, 16, 17.
    b. 2, 10, 13, 18, 21, 26.

c. 22, 23, 24, 25, 33.

d. 40.

Within the third or oreal group the degree of relationship is equal between 7, 15 and 27, so far as birds are concerned (cf. discussion of individual centres). For the mammals 7 and 15 appear as a unit, contrasted with 27. And for reptiles and amphibia, 7, 15 and 27 appear as a unit. This signifies that the degree of relationship between the oreal centres is conditioned by the very differing vagilities of the faunal elements adapted to the oreal environment.

Turning to the non-forest centres (group I), the most striking fact is the very close affinity of the Central American centres (Ia = 3, 4, 5). This would not be expected from the small-scale ecological differentiation of these centres. (cf. ELTON 1930, SAPPER 1937, HOLDRIDGE 1953, 1964, LAUER 1954, 1956, 1959, LÖTSCHERT 1955, CROIZAT 1958). The same can be said of the affinity groups Ib, Ic, Id and Ie. Once again, however, attention must be drawn to the close affinity between centres 27 and 39.

Within the arboreal group (II) the close connections within group IIb are very striking. The centres of this group lie above 14–1500 m. in the montane forest zone. The close affinities within this group indicate that the centres have all been controlled in their evolution by much the same factors.

Group IIa also makes a unit which can be contrasted with IIc, although there are a number of connections between IIa and IIc, some of which are very close (cf. discussion of individual centres). Within group IIc the affinities of 22, 23, 24 and 25 (i.e. the Amazonian centres), with each other are closer than with 33 (Serra do Mar centre). But if centres 22, 23, 24 and 25 are reckoned, for the purposes of argument, as subcentres of a big arboreal centre 'A', the closest relatives of the faunal elements of 'A' would be in centre 33.

The position of centre 40 (*Nothofagus* centre) is somewhat similar to that of 33. The high percentage of monotypic genera, especially in the amphibia, justifies separating centre 40 from all others. The closest connections of the *Nothofagus* centre are nevertheless indubitably with arboreal centre 'A'. Judging by the degree of distinctness of the faunal elements that occur in 'A' and 40, however, the affinity of these two centres is weaker than between 'A' and 33.

Ascribing centre 40 to group II contradicts the idea of a single Patagonian subregion which some authors have advocated e.g. HERSHKOVITZ (1969), FITTKAU (1969).

The close affinity of the centres within groups I, II and III expresses the close relationships of their faunal elements. This can be understood to mean that the phylogeny of non-forest, arboreal and oreal species has largely run separate in the three biomes, at least at the specific and subspecific levels of differentiation. It is appropriate to speak of the phylogeny of an arboreal, of a non-forest and of an oreal fauna.

It can thus be shown that at least 40 dispersal centres of general significance for the terrestrial vertebrate fauna exist in the Neotropical realm. But the level of differentiation of the faunal elements within these centres varies greatly. Monotypic genera, species of polytypic genera, and subspecies all occur together as elements of the same centre.

Two questions are therefore of great importance. Firstly, why do populations with such different levels of differentiation have ranges congruent with the same dispersal centre? And, secondly, how long can the species and subspecies by which a centre is recognised be considered as having been faunal elements of the centre?

From the viewpoint of present-day distribution, these problems can be approached mainly by considering polytypic, polycentric species. This is the case because subspecifically differentiated populations of a polycentric species are necessarily allopatric in distribution, for reasons of evolutionary genetics. MAYR (1967) said that: 'A subspecies is an aggregate of local populations of a species. It inhabits a geographical subregion of the range of the species and is taxonomically distinct from other populations of the species.'*

It is equally possible to approach these questions by considering supra-specific complexes, e.g. the distribution of the *Crax rubra* group in fig. 91. In such cases it must of course be clear at the outset that such a complex, made up of allopatrically distributed species, is a monophyletic group (cf. AMADON 1966, VUILLEUMIER 1965, MÜLLER 1972). It can be shown both for subspecies and for the species of a superspecies complex in MAYR's sense (1967) that the dispersal centre of which they are faunal elements must also have been their **centre of origin** (MÜLLER 1972).

These considerations are important because they show that the evolution of faunal elements that are differentiated as subspecies or as unit species of a superspecies, can be used to indicate the history of a distribution centre (MÜLLER 1972; fig. 98).

Each of the dispersal centres analysed here is also the centre of origin of certain subspecies, as shown in the discussion of the individual centres (cf. the subspeciation centres of *Crotalus durissus*). The geographical speciation and subspeciation found in most of the cases analysed is nothing but the: 'genetic alteration of a population during a period of geographical (i.e. spatial) isolation'** (MAYR 1967, p. 437). For this reason I consider that the dispersal centres arose because their faunas were isolated in them from time to time as refuge areas. This means, in the last analysis, that the dispersal centres of the Neo-

---

* 'Eine Subspezies ist ein Aggregat lokaler Populationen einer Art. Sie bewohnt eine geographische Unterregion des Verbreitungsgebietes der Spezies und ist taxonomisch von anderen Populationen der Art unterschieden.'
** 'genetische Umbau einer Population während einer Periode geographischer (räumlicher) Isolation'.

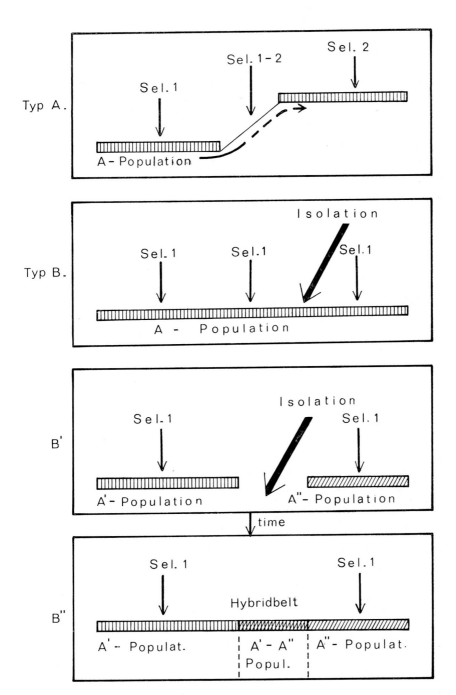

Fig. 98. The two general models of subspeciation (A and B). In A = differentiation as a result of migration and differential selection; in B = differentiation as a result of isolation.

tropical realm are the: 'centres of preservation of faunas and floras during regressive phases'* (DE LATTIN 1957, p. 402). The most recent period when a dispersal centre acted as a refuge in this manner can be worked out from the subspecific differentiation of the populations of a species which are confined to the centre by allopatry. Further discussion of the effects of spatial isolation can be found in: DARWIN (1845, 1859), WAGNER (1870, 1889), WALLACE (1880), MERTENS (1924, 1930, 1934), REINIG (1937, 1950), KOSSWIG (1942), MAYR (1942, 1967), DE LATTIN (1952, 1959), RILEY (1952) and LUDWIG (1954).

Species, and the individuals that make up species, have the power of dispersal. Furthermore populations, as soon as they are specifically distinct, can occur sympatrically alongside other specifically distinct populations. It is therefore possible for a species to have had its centre of origin in a totally different region from its present-day dispersal centre. The expansions and regressions of range which occur during the evolution of a taxon can lead, if big enough, to considerable displacements of range and even to complete geographical separation of the dispersal centre from the centre of origin.

When trying to work out, by using its subspecifically distinct faunal elements, the period when a dispersal centre last served as a refuge, it must first be recognised that subspeciation rate does not provide a reliable time scale. This is because the rate of subspeciation does not depend only on the effectiveness of isolation. It also depends on the gene pool of the population isolated, and the ecological diversity—niches and so forth—of the biotope (fig. 37). Moreover the actual moment when the last gene exchange occurred between isolates has only been worked out in a very few cases. When this moment of separation can be determined, it appears that the post-glacial period has been a long enough time for subspeciation in terrestrial vertebrates to occur.

A few examples will support this statement.

1. The house sparrow *Passer domesticus* was introduced into North America between 1852 and 1860. The North American populations are already subspecifically distinct (JOHNSTON 1969, 1970, JOHNSTON & SELANDER 1964, 1971, SELANDER & JOHNSTON 1967).

2. The British red deer, *Cervus elephas scoticus*, has been isolated from the central European populations since about 8000 years ago (MAYR 1967), and in recent times some populations of it have been introduced into New Zealand. The New Zealand populations already differ considerably in phenotype from the populations of origin.

3. According to CAMERON (1958) the island of Newfoundland in the Gulf of St. Lawrence became habitable by mammals at most 12000 years ago. Of the 14 species that occur there, ten form very easily diagnosed subspecies.

4. MARSHALL stated (1940) that the islands in the Great Salt Lake of Utah became isolated 8–10,000 years ago. The mammal species which occur on them are represented by endemic subspecies.

5. STEVEN (1953) showed that the *Clethrionomys* species (bank voles) of the

---

* 'Erhaltungszentren von Faunen und Floren während regressiver Phasen.'

184

Inner Hebrides are subspecifically distinct. They have probably been isolated for 7–9000 years.

6. MÜLLER (1966, 1968, 1969, 1970) reported on the vertebrate fauna of the islands of south-east Brazil which have been isolated for 6–11,000 years. A whole series of endemic subspecies occur on them among birds, mammals, reptiles and amphibia.

7. SICK (1967), HOGE (1966) and MÜLLER (1969, 1970) showed that subspecifically distinct populations were present in the campos of the island of Marajó in the Amazon delta. The island was completely inundated about 4000 years ago (BIGARELLA 1965, MÜLLER & SCHMITHÜSEN 1970).

These examples do not show that all subspecifically distinct populations must have arisen in the post-glacial period (cf. also DE LATTIN 1959, ANT 1964, MÜLLER 1972). They only show that subspeciation in many cases took place in the post-glacial. It would not be too misleading to assume, prima facie, that the dispersal centres last acted as preservation centres for their faunas and floras during regressive phases of these faunas and floras within the post-glacial period.

It is necessary to assume at the start that a regressive phase for the forest fauna will not be correlated with a regressive phase for the non-forest or oreal fauna. Indeed I consider that regressive phases for the forest fauna would coincide with expansive phases of the non-forest fauna. The position of the dispersal centres of the Neotropical land vertebrates therefore depended on the mode of action and history of Quaternary climatic oscillations and vegetational fluctuations. I shall argue this point of view in more detail in what follows.

Vegetational fluctuations during the Pleistocene of tropical Africa were established by EISENTRAUT (1968, 1970) and MOREAU (1963, 1966, 1969) by studying the comparative distribution of mammals and birds. Such fluctuations in vegetation had already been demanded by CHAPIN (1932) and MOREAU (1933) as necessary to explain the observed facts. They were of controlling importance for differentiation phenomena. GENTILLI (1949) and KEAST (1959, 1961, 1968) likewise suspected that a great part of the subspeciation of Australian taxa, and in GENTILLI's opinion a great part of their speciation also, was connected with the isolation of populations in small humid areas round the periphery of the continent during an arid period. These areas are the 'drought refuges' of MAYR (1967). The date ascribed to this dry period however, varies between 4000 and 20,000 years ago.

The pollen-analytical and geomorphological researches in Africa of BAKKER (1962, 1963, 1964, 1967), BOND (1963), COOKE (1964) and MORTELMANS & MONTEYNE (1962) have lent support to the theories of MOREAU (1963, 1966) and EISENTRAUT (1973). It is true that the arguments of MOREAU implied, perhaps rather prematurely, a close correlation between glacial and pluvial periods and between interglacial and arid periods. This was despite the fact that the problems involved have in no way been properly cleared up (cf. AUBRÉVILLE 1962, ROHDENBURG 1970).

CLARK (1962, 1963) and CLARK & FAGAN (1965) confirmed by C14 dating that the period of the latest arid phase in Africa was between 6800 and 3700 years ago This period is said to have led to vegetational fluctuations with considerable effects on the zoogeography. These vegetational fluctuations have furthermore been confirmed by palaeontological results in the Ethiopian realm (ARKELL 1964, BEUCHER 1963, BUTZER 1961, CAPOT-REY 1961, CLARK 1962, 1964, HUGOT, QUEZEL & MARTINEZ 1962, MONOD 1963 and MOREAU 1963).

AUBRÉVILLE (1962) and MOREAU (1966) both advocate the view that the montane forest biome shifted greatly in altitude because of the glacial periods. They also consider that the desert conditions that now exist in the whole Sahara region first developed only about 5000 years ago. MOREAU (1966, p. 59) stated that: 'The present extreme desert conditions throughout the Sahara were developed after about 5000 years ago.'

Vegetational fluctuations in the Ethiopian realm must obviously have been correlated with similar fluctuations in the Neotropical realm. It was geographers, palaeobotanists and limnologists who first drew attention to these phenomena (WAIBEL 1921, 1948, DUCKE & BLACK 1953, WILHELMY 1952, PAFFEN 1957, AUBRÉVILLE 1962, HUECK 1957, 1966, SIOLI 1968, STEVENSON & CHENG 1969, BIGARELLA 1971). It was only later that zoogeographers demanded vegetational displacements to explain the different types of geographical range and the evolution of individual taxa (VANZOLINI 1963, HAFFER 1967, 1969, MÜLLER 1968, 1969, 1970, 1971, 1972, MÜLLER & SCHMITHÜSEN 1970, VUILLEUMIER 1971).

The faunas of the Brazilian islands can be used to help date the displacement of the biochores in central and eastern South America (MÜLLER 1966, 1968, 1969, 1970, 1971, 1972). There is a whole series of islands which lie within the 50 m. isobath. Their age of isolation varies between 7000 and 11,000 years and can be worked out from eustatic variations in sea level in the post-glacial period, taking into account the differentiation rate of their fauna and tectonic processes which influenced sea levels locally (BIGARELLA 1965, BUISONJÉ 1964, FAIRBRIDGE 1958, 1960, 1961, 1962, 1967, FREY & EWING 1963, MÜLLER 1968). Out of all these islands I have analysed three zoogeographically and historically, paying special attention to the herpetofauna (MÜLLER 1968, 1970).

Of these three islands, Marajó is nearest to the equator, lying in the delta of the Amazon and Tocantins. It is also the biggest of the three, with an area of 47,964 km². Its greatest altitude is only 2 m. Thick rain forests on Quaternary soils, which are partly flooded during the rainy season, cover the western part of the island (BEEK & BRAMÃO 1968, SIOLI 1968). In the eastern part, open campo areas predominate. The island has been completely flooded at least twice by post-glacial rises of sea level (cf. fig. 99), and the fauna now to be found there could not have established itself until after these floods had subsided. The first such inundation happened between 5800 and 5200 years ago; this was during the Older Peron of FAIRBRIDGE (1962) or the Submergencia Alexandrense of BIGAREL-LA (1965) (cf. fig. 99). The second took place during the Younger Peron i.e. 4000 ± 150 years ago; this was during the Submergencia Cananeiense of BIGARELLA (1965). The amplitude of the rise in sea level in the Marajó region during the subsequent Submergencia Paranaguaense is not yet known. According to HURT (1964) the age of the Submergencia Paranaguaense is 2300 ± 220 years, on the evidence of C14 dating.

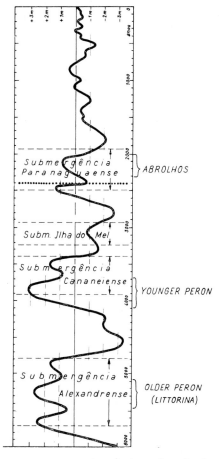

Fig. 99. The sequence of post-glacial eustatic variations of sea level on the Brazilian coast (after FAIRBRIDGE 1962, BIGARELLA 1965). The horizontal arrow indicates the end of the post-glacial arid phase on the Curitiba plateau.

The island of São Sebastião lies on the tropic of Capricorn, is 350 km² in area and reaches a height of 1400 m. Its orogeny can be correlated with that of the Brazilian coastal range. Like Marajó it is screened off by thick rain forest from the campo and caatinga regions which run transversely across central Brazil.

The island of Florianopolis, which reaches a height of 540 m., is the southernmost of the three that were specially studied and lies off the coast of the state of Santa Catarina. It was divided into at least five smaller islands by eustatic rises in sea level during the Submergencia Cananeiense. In geological structure and geomorphology it is very like the south Brazilian coastal range. Marine

187

sediments and present-day dune formation in the interior of the island indicate that, during the post-glacial transgression it must, like Marajó, have been smaller in extent than now. Also it was probably connected with the mainland 4–5000 years ago, in the interval between the Submergencia Cananeiense and the Submergencia Paranaguaense. (BIGARELLA 1965, DELANEY 1963, 1966, HURT 1964, cf. fig. 96).

The three islands lie within the natural limits of the South American rain forests (Serra do Mar, Para and Guyana centres). But the obvious expectation, that they would have a pure forest fauna, is only fulfilled for São Sebastião. It is likely that the herpetofauna of this island is known completely (MÜLLER 1968). Apart from the introduced form *Hemidactylus mabouia* (MÜLLER 1969) and three sea turtles that are unsuited for zoogeographical analysis (i.e. *Chelonia mydas*, *Eretmochelys imbricata* and *Dermochelys coriacea*), São Sebastião is inhabited by a pure forest fauna with three different types of geographical range. The first, polycentric type, comprises species whose range includes the whole of South America of tropical character. Such forms also occur in the open campo areas of Brazil, inside the gallery forests. The second type includes monocentric faunal elements of the Serra do Mar centre. And the third and most interesting type has a polycentric range with an Amazon centre and a Serra do Mar centre separated by the central Brazilian campo and caatinga regions; this type of distribution is also discussed under the Serra do Mar centre. Among flightless groups—i.e. amphibia, reptiles and mammals other than bats, non-forest species are completely lacking on São Sebastião, although the present-day ecology would allow them to exist (cf. MÜLLER 1968).

On Florianopolis 72% of the herpetofauna can be assigned to these same three distribution types. But here there is yet another type represented by Uruguayan faunal elements. These obviously reached the island from the mainland, coming up from the south before the island was isolated. The same may also have happened on some of the islands that lie east of Florianopolis. Thus in 1967 I was able to show the presence on the little island of Campeche of *Leptodactylus gracilis*, which must be reckoned as a Uruguayan element. The population on Campeche is subspecifically distinct from that of Floriano-polis or the mainland with the name *Leptodactylus gracilis delattini* (cf. MÜLLER 1968).

On Marajó, Uruguayan elements are lacking just as on São Sebastião. How-ever in the eastern campo zone I have found species, some of them in the form of subspecifically distinct populations, which are typical representatives of the Neotropical campo formation of the Caatinga, Campo Cerrado and Caribbean centres. This type of distribution is represented, for example, by *Crotalus durissus* (with the endemic subspecies *C. d. marajoensis*), *Coryophaspiza melanotis* (with the endemic subspecies *C. m. marajoara*), *Bothrops marajoensis*, *Cyclagras gigas*, *Aratinga cactorum*, *Cnemidophorus lemniscatus* [which is widely distributed in open landscapes of northern South America and belongs to the ecophysiological type of 'desert lizards' according to MÜLLER (1971)] and *Myrmecophaga tridactyla*. The distinct Marajó species are also represented in the isolated 'campo islands' in the rain forest between Santarem and Marajó (cf. fig. 5 in MÜLLER 1970).

The individual types of distribution and the degree of differentiation of the populations only make sense in terms of post-glacial displacements of the bio-chores (cf. MÜLLER 1970, MÜLLER & SCHMITHÜSEN 1970). This means that during and after the time of the post-glacial transgression on Marajó there was an expansion of the campo, corresponding to the Ilha do Mel submergence (fig. 99). During the course of this expansion Campo Cerrado elements, among others, reached Marajó. This phase of campo expansion gave way later to a phase when the forests expanded, corresponding to the Paranagua submergence. This expansion of forests broke up the non-forest migration corridors (for the older history of Marajó see LUDWIG 1966, 1968).

These successive expansions of campo and forest are demanded in the first instance on zoogeographical grounds. They must also have been important for the Para, Guayana, Madeira, Serra do Mar, Caatinga, Campo Cerrado, Uruguayan and Parana centres. This raises the questions whether these zoo-geographical results are of general significance and whether they can be con-firmed by the results of other disciplines, such as geology, geography and palaeontology.

In point of fact a dry phase can be established in South America for the period between 5000 and 2300 years ago i.e. from the Alexandra submergence to the Paranagua submergence (cf. fig. 99). This led to an expansion of the campo areas, in the broad sense, and a regression of the rain forests (AB'SABER 1962, 1965, 1970, BIGARELLA 1965, PIMIENTA 1958, TRICART, VOGT & GOMES 1960, MÜLLER & SCHMITHÜSEN 1970, HAFFER 1970, VANZOLINI 1970, VANZOLINI & WILLIAMS 1970). During this phase, henceforth called the post-glacial arid phase, there was a spreading out of the restinga from Cabo Frio to Rio Grande do Sul (DANSEREAU 1947, DELANEY 1963, 1966). On the basis of C14 dating of mollusc shells in dead dunes, HURT (1964) has dated the expansion phase of the restinga as between 4500 and 2500 years ago.

During the post-glacial arid phase a continuous strip of restinga can be shown to have existed between Rio Grande do Sul and Paranagua and Cananea on the one hand, and between Ilha Grande (Rio de Janeiro) and Cabo Frio on the other (PIMIENTA 1958, TRICART, VOGT & GOMES 1960). At least in the south this must have acted as a migration corridor for Uruguayan faunal elements (MÜLLER 1970). The island of Florianopolis is now separated from the mainland only by a channel 2 m. deep and Uruguayan faunal elements reached it during this arid phase. Compare the oscillations in sea level in the interval between the Alexandra and Paranagua submergence in fig. 99.

The non-forest species which now occur isolated in the high campos of the Serra do Mar (cf. fig. 73) widened their ranges during this arid phase and their populations were in gene exchange with each other (AB'SABER 1965, MÜLLER 1970). In the same arid phase there originated the dune formations and shell mounds which are now to be found in the rain-forest regions of Rio Grande do Sul, Santa Catarina, Florianopolis and Parana (HURT 1964, VANZOLINI & AB'SABER 1968).

VANZOLINI & AB'SABER (1968) have considered the populations of *Liolaemus lutzae* (see MERTENS 1938) which are now isolated in the restinga of Guanabara. They take the view that these lizards separated off from the parent group *Liolaemus occipitalis* in Rio Grande do Sul during the post-glacial arid phase. They then wandered north by way of the expanded restinga zone and were isolated by the subsequent forest expansion at the time of the Paranagua submergence (cf. fig. 99). The existence of such a corridor of migration, however, is only proved as yet for the states of Rio Grande do Sul, Santa Catarina and Parana (MÜLLER 1970). As a feature of any general significance it could not have extended up to the state of Rio de Janeiro since only a single species managed to move northwards along it as far as Cabo Frio. And furthermore the island of São Sebastião has a pure forest fauna (MÜLLER 1968) and there are no other isolated *Liolaemus* populations in the coastal campo between São Paulo and the northern part of Santa Catarina.

The phase of expansion of the campo and restinga areas represented a phase of regression for the rain forest (BIGARELLA & ANDRADE 1965). The division of the Serra do Mar centre into subcentres fits this view completely for there was an expansion of open landscape during the post-glacial arid phase both between the Paulista and Bahia subcentres in the Cabo Frio region and between the Bahia and Pernambuco subcentres in the Salvador region.

TRICART, SANTOS, SILVA & SILVA (1958) reconstructed the most recent phase of forest regression in Bahia. They became convinced that, apart from a small coastal forest in the north-east (i.e. the Pernambuco subcentre) the whole southern part of the region that they had worked on, including the coastal zone, had been considerably drier than at present, though with episodes of very intense torrential rain and mud flows.

It is possible to suspect, though not to prove, that the Serra do Mar subcentres were also isolated from each other earlier than the post-glacial arid phase, about at the end of the Würm glacial. Obvious signs of glaciation in the Serra de Mantiqueira and on the Pico de Bandeira (MAACK 1958) indicate a cooler climate in the Würm. This also affected the lowlands, as in Colombia (cf. WILHELMY 1957), and must have led to shifts in altitude of the biota.

During the post-glacial arid phase the campo cerrado and caatinga did not expand only to the south and south-east, but also to the north, as the fauna of Marajó shows (MÜLLER 1968, 1970, HAFFER 1969). The populations of the hyleal 'campo islands', which are isolated at the present day, would then have been in genetic contact with the Campo Cerrado populations.

REINKE (1962) was able to show that dry regions of the Venezuelan Llanos were connected with the campo regions of central Brazil by means of a corridor of savanna country, running by way of the Amazonian campo islands (cf. fig. 34). In this corridor, or 'climatic bridge', the rain forest competed for space with the campo (cf. KOEGEL 1922, SIOLI 1968). The available zoogeographical results fit perfectly into REINKE's picture. The populations of the hyleal 'campo islands', which are situated in the area of this former corridor of savanna and which have a fauna and flora specific to the campo, are relics of a migration route for campo species connecting north and central South America during the post-glacial arid phase. This explains why there are no arboreal dispersal centres between the Xingu and the Tocantins and why the limit between the Para and Madeira centres runs just here (cf. hybridbelt in fig. 98). Arboreal centres are also absent in part of the region between the rivers Xingu and the Tapajoz. Thus the present-day distribution of the campo monkey *Callithrix argentata* demands a migration corridor by way of the campo islands between these rivers. The subspecies of this species living in this region are confined to the campo islands (cf. fig. 92).

The expansion of campo in the post-glacial arid phase has also been important for the separation of the Amazon and Madeira centres, in addition to the isolating effect of the Rio Madeira (HAFFER 1968, MÜLLER 1970). The occurrence of *Crotalus durissus* in the isolated campo islands west of the towns of Porto Velho and São Antonio on the Rio Madeira (MÜLLER 1968) indicates that these campo islands are natural and also that they have been connected in the post-glacial period with the campo cerrado of the Serra dos Parecis.

The extent of the campo expansions in the south-west part of the Amazon centre is not yet known with certainty. The birds of the unforested Urubamba valley, between the Marañon centre and the campo areas of Bolivia, total 66 species (CHAPMAN 1921). Thirty-eight of these are widespread species that prefer or require an unforested biotope; only 19 species arose without doubt from the Campo Cerrado centre. The degree of differentiation suggests that in the post-glacial period the Urubamba valley was not so distinctly separated by rain forest from the open landscapes of Bolivia and central Brazil as it now is.

The populations of *Crotalus durissus* in the campos of Obidos are closely related to those of the savanna of Paru and of the Roraima centre. This suggests that during the post-glacial arid phase there were also expansions of campo between the Rio Maicuru and Rio Cumina (cf. fig. 6 in MÜLLER 1970 and the course of the savanna corridor in fig. 34).

Crotalus durissus marajoensis is closest related to the *C. durissus* populations in the coastal savanna of the Guyanas and the Caatinga centre. This shows that corridors for the migration of non-forest species must also have existed along the coast by way of the campos of Amapa (fig. 100). The existence of the endemic subspecies *Crotalus durissus dryinus* in the savannas of the Guyanas also indicates that these savannas are natural (cf. BAKKER 1954, GOODLAND 1966, WYMSTRA & HAMMEN 1967).

The position of the dispersal centres in north-west South America and the faunal elements within them also demand post-glacial fluctuations in vegetation. The non-forest Cauca and Magdalena centres, as already explained, received their faunal elements from the Barranquilla centre. But at the present day the arboreal Nechi subcentre forms an isolation barrier between these centres (CUATRECASAS 1958). In the opinion of CHAPMAN (1917) and HAFFER (1967, 1970) it can be taken as certain that the Nechi subcentre was easier to pass through during some 'interglacial' arid phase. This allowed an exchange between the Barranquilla centre and the inter-Andean centres. The lack of endemic species in the Magdalena centre, however, makes it obvious that this centre has not for long

Fig. 100. Migration routes of faunal elements of non-forest centres during the post-glacial arid phase (after MÜLLER 1970, with additions and modifications).

been isolated from the Barranquilla centre. It is likely that the Cauca and Magdalena centres were in ontact with the Barranquilla centre during the post-glacial arid phase when dunes were forming in the Llanos of Colombia and Venezuela (GOOSEN 1964, BLYNDENSTEIN 1967). The isolating power of the relict rain forests in the middle Magdalena graben during this arid phase is not known.

As concerns the Choco and Nechi faunal elements, their divergent evolution also requires displacement of the biomes. Several authors have already demanded vegetational fluctuations in these areas on the basis of geographical, pollen-analytical and faunistic studies (WILHELMY 1954, 1957, HAMMEN 1961, HAMMEN &

Plate 1. The rain-forest biom (Island of Florianopolis, Santa Catarina, Brazil, march 1969).

Plate 2. The Campo cerrado biom (near Pousada do Rio Quente, Goias, Brazil, march 1969).

193

GONZALES 1960, HAFFER 1967, 1970). The researches of WEST (1957) on the Rio Atrato and GOOSEN (1964) in the Llanos of eastern Colombia indicate that, during the post-glacial arid phase, there was an intensified formation of savannas in the Atrato region (WEST 1957, map II) and also marine transgressions (cf. Surface Geology in WEST 1957, p. 17, BEDERKE & WUNDERLICH 1968, p. 532).

The faunal elements of the Barranquilla centre are at most subspecifically distinct. This suggests that the ancestors of the non-forest forms now found in the centre did not reach it until the post-glacial period, any previous invasions of such forms into the centre having died out. The Barranquilla centre probably represents a 'filter region' which alternated between a more arboreal and a more non-forest state. The centre certainly served at very different times in the Quaternary as an immigration route for non-forest species from North America to South America, and vice versa (cf. the detailed discussion in STIRTON 1950). This is made obvious by the fact that the Barranquilla populations of *Crotalus durissus* are closer related to those of the Central American Pacific centre than to those of the Caribbean centre (HOGE 1966).

STIRTON (1950) was the first to realise that SIMPSON's concept (1950) of intercontinental faunal exchange between North and South America could only be made to work if correlated fluctuations in vegetation had occurred (see also BENNETT 1968).

The works of AUER (1960) and LANNING (1965, 1967) show that the *Nothofagus* and Andean Pacific centres also underwent at least marginal expansions and regressions during the post-glacial arid period. It is not precisely known when the expansion phase of the Andean Pacific centre began. But it has been established by C14 dating of the remains of extinct forests on the Peruvian coast that about 8500 years ago the forest margins of the Choco subcentre lay a1 least 400 km. farther south than they do now (KOEPCKE 1961, LANNING 1965, 1967). Displacements of the mangrove vegetational formation were correlated with these fluctuations of the forest (SCHWEIGGER 1959, KOEPCKE 1961). Thus *Arca grandis*, a Pacific bivalve mollusc that inhabits mangroves, now reaches its southern limit at Tumbes, but 8000 years ago extended as far as Sechura (LANNING 1965).

In considering the true age of the north-western centres of South America, as opposed to the latest period when they acted as refuges, the origin of the northern Andes is relevant. These mountains began as discontinuous chains of islands in the course of the Tertiary. These chains gradually grew more continuous, but remained separated from each other by narrow, rapidly sinking sedimentation troughs—the sites of the Cauca and Magdalena centres (WOODRING 1966). The last strong folding happened in the middle or upper Pliocene, but the height of the mountains could not have been very great at that time. This is shown by the post-orogenic, upper Pliocene, fresh-water Tilata Beds of the Bogotá basin which rest discordantly on Cretaceous and Tertiary. At the present day these beds lie at a height of 2600 m. but they contain fossil seeds and pollen of a pure, tropical lowland flora (HUBACH 1958, HAMMEN 1966, HAMMEN & GONZALES 1964). As against this, the overlying Sabana Formation, of Pleistocene and Holocene age, exclusively contains pollen of an upland flora (HAMMEN & GONZALES 1964). This justifies the conclusion that the elevation of the eastern Cordillera, as opposed to its folding, did not happen until the end of the Pliocene or the beginning of the Pleistocene. This elevation must have caused a decisive change in the local climate of the region, and naturally also in the vegetation. These geological considerations, however, do not permit a general dating of Choco faunal elements, as explained above in discussing the Choco subcentre. There is evidence, for example, of immigration at more than one period of time.

The position of Central American distribution centres has also been influenced by vegetational fluctuations during the Quaternary (STIRTON 1950, MARTIN 1958, DUELLMAN 1960, MAYR 1964). There are isolated occurrences of Central American Pacific faunal elements in the peninsula of Azuero Santos, some of them as subspecifically distinct populations i.e. *Crypturellus cinnamomeus praepes*, *Emberizoides herbicola lucaris*, *Aimophilia rufescens hypaethrus*, *Arremonops rufivirgata superciliosa*. An unforested thoroughfare along the Pacific coast is required to explain these occurrences at least for a short time.

The different numbers of species in the Mosquito and Chiriqui subcentres of the Costa Rican centre support the view that vegetational fluctuations have been important. The Mosquito subcentre is richer in species than the Chiriqui subcentre probably because the latter has been more subject to fluctuations in its vegetation.

The close relationship between the Yucatan and Central American Pacific centre also makes sense only if an expansion of savanna had occurred during the Quaternary in the region of the plain of Tehuantepec.

The expansion of open landscapes during the post-glacial arid phase that has just been discussed (cf. fig. 100), gave way in the plateau of Curitiba (Brazil) to a phase of forest expansion. This started at the beginning of the Paraguana submergence i.e. about 2400 years ago (HURT 1964, AB'SABER 1965, BIGARELLA 1965, cf. fig. 92). This expansion of forests has continued to the present day (RAMBO 1946, 1948, 1956, LAUER 1952, WILHELMY 1952, PAFFEN 1957, MAACK 1962, DUELLMAN 1960). It has been recorded in both Central and South America. Exceptions were recorded in Patagonia by AUER (1933, 1960). The results of WILHELMY (1952), PAFFEN (1957) and MAACK (1962) agree with my own observations.

The non-forest corridors through or towards Amazonia were disrupted during this phase of forest expansion. Compare in this connection the ranges of *Monodelphis domestica, Picumnus varzeae, Phaetornis nattereri, Callithrix argentata, Crotalus durissus, Coryophaspiza melanotis, Aratinga cactorum.* The hyleal 'campo islands' became isolated. Some of them may owe it to human activity that they have not been completely overgrown by forest; the work of ZERRIES (1968) on the development of the Indian cultures in the Amazon is relevant in this connection.

The Chaco, Campo Cerrado and Caatinga centres, at least in the region of gallery forests, became more arboreal than they were during the post-glacial arid phase. But, except in a few cases, the latest phase of forest expansion did not lead to gene exchange between the faunal elements of the Serra do Mar centre, on the one hand, and the Amazon, Madeira and Para centres on the other.

This appears from the distribution of 97 species of bird whose ranges include the Serra do Mar centre and the rain forest centres of Amazonia, but which are split in two by the Campo Cerrado and Caatinga centres. The Serra do Mar populations almost always form clearly distinct subspecies in these cases. There are only three exceptions, as discussed under the Serra do Mar centre.

The same disjunction is found in the ranges of many genera. Compare in this respect the ranges of *Xipholena, Procnias, Cotinga, Drymophila, Haplospiza, Amaurospiza, Pitylus, Chiroxiphia* and *Schiffornis*. The amphibia, reptiles and mammals show the same disjunction as the birds (cf. WILLIAMS & VANZOLINI 1966, MÜLLER 1968, 1970).

The level of differentiation of these disjunct populations leads inescapably to the conclusion that **the isolation of the Amazonian rain forests from those of the Serra do Mar is older than that between the campo cerrado and the hyleal campo islands.** This indicates furthermore that the campo cerrado is 'natural' as WAIBEL already insisted in 1921 and as is confirmed by the existence of the Campo Cerrado, Caatinga and Chaco non-forest dispersal centres. Pleistocene fossils, such as *Chrysocyon brachyurus* found at Lagoa Santa in Minas Gerais, show that the campo cerrado already existed as a vegetational formation in the Pleistocene, as WILHELMY (1952) supposed.

The obvious barrier-line for typical non-forest species in the region of Tehuantepec in my opinion is not due to pre-glacial tectonics as STUART (1966) supposed. It is rather the result of a strong expansion of the forest area after the post-glacial arid phase (DUELLMAN 1960). During this time of forest expansion some species of the Central American rain-forest centre must have succeeded in immigrating again into the Pacific side of Central America. They are now to be found in isolated relict forests in El Salvador (EMSLEY 1964, MERTENS 1952).

However it is possible that some of the populations in these relict forests came in by way of the Chiapas valley west of Mount Tajumulco. STUART (1954) and CARR (1950) have described this corridor and DUELLMAN (1966) has drawn attention to its importance for eurytopic South American immigrants such as *Constrictor constrictor*. Compare also *Tantilla armillata* as discussed by HARDY & COLE (1967). STUART's supposition (1954), that the disjunction of the populations north-east and south-west of the Coastal Range happened in the Pliocene is not likely, since the isolated populations are only very weakly distinct. No more likely is the view that the distribution of Central American vertebrates indicates a pre-glacial marine connection between the Pacific and Atlantic by way of the plain of Tehuantepec, as suggested by BURT (1931), GLOYD (1940), OLIVER (1948) and STUART (1941). Such a marine connection would also contradict the geological and palaeontological results of DURHAM, ARELLANO & PECK (1952, 1955), OLSON & MCGREW (1941) and STIRTON (1954).

The post-glacial arid phase, which gave way to a phase of forest expansion about 2400 years ago, was also preceded by a phase of forest expansion. This was correlated with a pluvial phase that marked the end of the Würm glacial (WILHELMY 1957, HAMMEN 1966). This earlier phase of forest expansion must have been very important for the montane forest centres (nos. 2, 13, 21 and 26) which, compared with the lowland centres, are extremely rich in endemics. The faunal elements of the montane forest centres come without exception from the zone above 1400–1500 m.

The present-day dispersal rate of the montane forest fauna is small (cf. MAYR & PHELPS 1955, 1967). Furthermore the subspecific distinctness of isolated montane forest populations, as for example in the Pantepui centre, shows that montane forest forms cannot pass through lowland rain forest to any considerable extent. The differentiation of the montane forest fauna nevertheless shows clearly that the dispersal rate must once have been much greater.

Some authors have argued that now-isolated montane forest populations can be explained if mountains at present separate were originally connected together by highland (TODD & CARRIKER 1922, CHAPMAN 1917, 1926). Thus the highlands of Talamanca are supposed to have been connected with the Western Cordillera, the Sierra de Santa Marta with the Eastern Cordillera and even the Pantepui region connected with the Andes.

GRISCOM (1932), on the other hand, was the first to advocate the view that during the Pleistocene cooling: 'the avifauna of the Subtropical Zone in Central America descended to sea-level and had consequently a chance to pass continuously from Mexico to Colombia'. HAFFER (1967) also adopted GRISCOM's view and DUELLMAN (1960, p. 45) insisted that: 'in southern Mexico and northern Central America climatic fluctuation during the Pleistocene was of sufficient magnitude to cause vegetational shifts, both vertically and latitudinally, resulting in the establishment of alternating continuous and discontinuous lowland and highland environments, although this climatic fluctuation was not so great as to eliminate tropical lowland environments from the region'.

The work of van der HAMMEN & GONZALES (1960) showed that during the Pleistocene of eastern Colombia a subtropical climate predominated at a height of only 500 to 700 m. This result was confirmed by the work of WEYL (1955) in Costa Rica, by MONROE (1968) in Honduras, by HUTCHINSON, PATRICK & DEEVEY (1956), SEARS, FOREMAN & CLISBY (1955), MARTIN & HARRELL (1957), MARTIN (1958), WHITE (1956) and DORF (1959) in southern Mexico, by WILHELMY (1957) in northern South America and by MOREAU (1966) in Africa. Moreau wrote (1966, p. 59) that: 'To summarize the most salient features of the ecological vicissitudes of the Pleistocene, it can be said that during the glaciations a continuous block from Abyssinia to South Africa had a montane climate..... This situation was fully developed for the last time from about 25,000 to 18,000 years ago, but during most of the last 70,000 years the ecological picture presented by Africa has been nearer to that associated with the glacial maximum than to that of the present day; and the balance between montane and lowland that we now see in Africa is the result of changes between about 16,000 and 8000 years ago.'

The sea level on the east coast of South America 11,000 ± 150 years ago was about 115 m. lower than at present (FRAY & EWING 1963, BIGARELLA 1965). Between 7327 ± 1300 and 7803 ± 150 years ago the sea level in the Bay of Santos (Brazil) was still 30 m. lower than at present (BIGARELLA 1965, HURT 1964). On the other hand about 5800 years ago a strong marine transgression is indicated in south-east Brazil. This is the Submergencia Alexandrense of BIGARELLA (1965) and the Older Peron or Littorina period of FAIRBRIDGE (1962).

I therefore believe that about 11–12000 years ago the montane forest biome still extended down to lower altitudes than at present and that montane forest species partly lived in what are now lowland forest dispersal centres. As the temperature began to rise the altitudinal range of the montane forest fauna shifted uphill. This necessarily led to the isolation of montane-forest populations.

The faunal elements of the Santa Marta centre show a lower degree of differentiation than those of the Sierra Nevada centre which suggests that most Santa Marta elements are younger. This difference in degree of differentiation can be explained by supposing that, because of changes in the altitudinal ranges of the fauna, the Sierra Nevada centre used to lie 11,000 years ago where the Santa Marta centre is now—according to MACHATSCHEK (1955) there are indications of glaciation in the Sierra Nevada. Only when warming started to take place, and the biota shifted uphill, did the Santa Marta centre become available to lowland forest species.

The work of GRISCOM (1932), HAFFER (1967), SKUTCH (1967) and MONROE (1968) has made it likely that during the Pleistocene the altitudinal range of highland birds in Central America and northern Colombia was greater than now and the basimontane biome predominated. In contrast to the view of MAYR & PHELPS (1955, 1967) I consider that these relationships also hold for the Pantepui centre.

The situation of the montane forest centres and the affinities between them can therefore be explained by supposing that, about 11,000 years ago and earlier, the true range of action of the forest fauna was greater than now and that isolation in the montane forest biome set in afterwards. The beginning of this isolation may have been correlated in time with the younger Dryas period (cf. BUTZER 1957, CHANG 1969, FIRBAS 1961, FRENZEL 1968, LINDROTH 1970, WOLDSTEDT 1958). It is not yet known precisely how the warm Allerød oscillation, that can be proved in Europe, affected South America. Even in Europe the changes in range that are due to it are by no means fully understood (COOPE 1969, LINDROTH 1970).

MOREAU's opinion (1966) that most of the species of the montane forest biome of Africa are only 8000 years old, in accordance with the time when isolation began, does not seem to me generally applicable. On the contrary, it is likely that the populations belonging to a polycentric polytypic species that now occur in different montane forest centres only began to differentiate **subspecifically** from each other about 8000 years ago. Compare in this connection the problem of allopatric differentiation (MÜLLER 1972).

Given that the montane forest biome extended farther downhill 11,000 years ago, the fauna of the lowland rain forests must also have shown displacements in its dispersal centres compared with the present. In the present state of knowledge, however, it is not possible to say what these were.

The zoogeographical results arrived at here make it likely that before the post-glacial arid phase the campo cerrado, in edaphically suitable places, was more arboreal than it now is. This is likely because an exchange between Amazon and Serra do Mar faunal elements would only be possible by way of enlarged islands of forest or enlarged gallery forests (Rio Araguaia, Rio Tocantins). Such exchange is necessary to explain the Amazon–Serra do Mar type of distribution.

The relationships of the oreal fauna are only partly comparable with those of the montane forest fauna, except as concerns the time when isolation began in the post-glacial. Close connections do exist between the oreal and the lowland biomes, but nonetheless affinities in a vertical direction are weaker than in a horizontal direction. This is not only true for terrestrial vertebrates—compare the ranges of the plant genera *Salix* and *Alnus* in SCHMUCKER (1942), and WEBER (1958), and also the works of BADER (1958), ESPINAL, LUIS & MONTENEGRO (1963) and FRANKLIN (1912). The predominance of horizontal over vertical affinities is illustrated by comparing the Puna centre with the Patagonian centre and the North Andean with the Talamanca paramo centre. Also interesting in this connection is the distribution of the marsupial family of Caenolestidae. This consists of three genera: *Caenolestes*, *Lestoros* and *Rhyncholestes* with seven species in all. The five *Caenolestes* species—*C. caniventer*, *C. convelatus*, *C. fuliginosus*, *C. obscurus* and *C. tatei*, live in the montane forest biomes and paramos of Ecuador, Colombia and eastern Venezuela. The genus *Lestoros* is a Puna faunal element and *Rhyncholestes* is a *Nothofagus* faunal element. Moreover it appears that the number of species and subspecies with very restricted ranges is extremely high in the Central Andes of Peru. Towards the south, however, it tends asymptotically to zero. In the oreal biome of Chile vertebrates do not occur as indicators, being replaced by invertebrates (MANN, 1968). Thus there are only eleven species of mammal in the oreal biome of Chile and Argentina (CABRERA 1957–1961). All of them are believed to have come in relatively recently from the north. This is parallel to the way in which the alpine 'massifs de refuge' were populated, for in the strict sense these massifs are only significant for invertebrates and the present-day vertebrate fauna consists of young, post-glacial immigrants (BEAUMONT 1968, BESUCHET 1968, NADIG 1968, SAUTER 1968).

The lack of vertebrates among the faunal elements of the oreal biome of Chile is probably due to an intense glacial phase in the southern Andes. During this, the oreal biome in most cases was completely covered with ice and the only animals that could live there belonged to the marginal snow fauna ('Schneerandfauna' of KÜHNELT 1969). This supposition is supported by results from glacial geology (AUER 1961, CALDENIUS 1932, HASTENRATH 1967, MARTIN 1965, MERCER 1965, 1968, PUTZER 1968, VIERS 1965, POLANSKI 1965, WEISCHET 1964). It is likely that during the Würm glacial no strip of treeless ground remained between the edge of the ice and the forest. The lower temperature did not lead to a mere vertical displacement of the altitudinal zones of vegetation as these now exist, but the forest biome extended right up to the edge of the ice.

As a consequence the Andean region of South and Central America has only three clear dispersal centres for terrestrial vertebrates. These show strong orographic subdivision and consequently can be divided not only into sub-centres, but into even smaller tertiary centres.

The two oreal centres within South America partly agree with the regional subdivision of the Andes given by MANN (1968). Thus the North Andean centre corresponds to MANN's 'nordandine Gemeinschaft' and the Puna centre to MANN's 'Hochplateaugemeinschaft'. MANN's 'Südandengemeinschaft' cannot be recognised from the vertebrates for the reasons already explained, but funda-mentally this does not disagree with MANN's division of the South American Andes into three regions.

The presence of an endemic Puna fauna and the existence of a number of endemics in Lake Titicaca argues against the view that the Puna basin was completely glaciated during the Würm glacial.

One Neotropical dispersal centre was only indirectly influenced by fluctu-ations in the vegetation. This was the Galapagos centre. Its fauna can be divided into two groups according to origin. The more strongly differentiated group points to connections with Central America and the Antilles while the less differentiated group is connected with the Andean Pacific centre. SCHWEIGGER (1959) considered that the Galapagos must once have been connected by land with South America. This however seems quite unnecessary precisely because of the different degrees of differentiation of the groups of origin, along with the known geological evolution of the Galapagos (cf. PUTZER 1968) and consider-ation of what species are actually present. Thus the lack of amphibia is striking. I conclude that the fauna of the Galapagos centre probably came in by sea, with different relationships of the marine currents in the Pacific off South America for the two groups of immigrants before and during the Pleistocene. WILHELMY (1957) believes that the Peru current has scarcely changed its course since the Miocene. It seems to me, however, that during the earlier part of this time the currents may have been different from what they now are because of the connection between the Caribbean Sea and the Pacific through Panama which lasted at least into the Pliocene (THENIUS 1959, HERSHKOVITZ 1969). This explains why the older parts of the island fauna have Central American affinities. After the closing of the Panama bridge (Bolivar geosyncline) the present pattern of currents was established and Andean Pacific species were able to populate the Galapagos.

Finally I should like to discuss **the significance of dispersal centres for the evolution of Neotropical taxa and also from the point of view of geographical studies.**

DE LATTIN (1957) has already pointed out the importance of dispersal centres for evolution in general. The Neotropical centres, however, have a special significance of their own. They are a key to understanding why species are so numerous in tropical biomes (cf. MULLER 1972).

Recent discussions of this problem (DOBSHANSKY 1959, MOYNIHAN 1962, HUTCHINSON 1965, VUILLEUMIER 1969, MACARTHUR 1969), have assumed that favourable climatic conditions, an increased rate of mutation, heightened competition and a potentially large number of niches have allowed rapid evolution in the Neotropical rain-forest biome.

On the other hand HAFFER (1967, 1969, 1970) and MÜLLER (1968, 1970, 1972) have advocated the view that most of the herpetofaunas and avifaunas of Amazonia owe their divergent differentiation to phases of isolation in forest areas smaller than what now exist, during expansions of the campo. MÜLLER (1968, p. 60) wrote: 'Intense splitting up of forest must have occurred in the younger Tertiary and during the interpluvial periods. This would have increased to an extraordinary extent the degree of isolation of populations of species strictly adapted to rain forest, and without doubt is one of the reasons for the richness of the Neotropical rain forest in species.'* HAFFER (1969, p. 131) wrote: 'Most species probably originated in forest refuges during dry climatic periods.'

The fact that dispersal centres also exist inside the Amazonian rain forest is an argument for the views of HAFFER (1969) and MÜLLER (1968, 1972).

Every time that a dispersal centre functions as a refuge during regressive phases of the environment that it represents, one of the most important evolutionary factors starts to act; this is the spatial isolation of parts of originally uniform populations. This spatial isolation is an, or perhaps the, important precondition for divergent evolution. I therefore follow DE LATTIN (1957, p. 406) who considered that the dispersal centres of Holarctic land animals were: 'centres of evolution for species and subspecies'.** What he asserted also holds for the Neotropical realm.

I have been able to show in this book that most subspecifically distinct populations of polycentric species are monocentric for one of the Neotropical dispersal centres. Island subspecies are an exception, as shown by the centres of subspeciation of *Crotalus durissus*. But subspecies can be considered as species *in statu nascendi* (MAYR 1967 etc.) although, as already explained, they have to be allopatric in dispersal for reasons of evolutionary genetics. From this point of view therefore, a dispersal centre can also be taken as the centre of origin for species. If a distribution centre acts for long enough as a refuge, then subspeciation leads to speciation, if the populations do not first die out. The length of time necessary for a subspecies to become a species depends both on external and internal factors, as also does the extinction rate of populations (cf. MACARTHUR & WILSON 1967, WILSON & SIMBERLOFF 1969).

Two external factors which raise extinction rates abruptly are strong oscillations in climate and the sudden appearance of competitors. Available studies in geomorphology and glacial geology indicate that the climatic conditions in

---

* Eine starke Waldzersplitterung, die während des jüngeren Tertiärs und während der Interpluvialia bestanden haben muß und die den Isolationsgrad der Populationen ökologisch streng an den Regenwald angepaßter Arten außerordentlich erhöhte, ist zweifellos mit ein Grund für den Artenreichtum des neotropischen Regenwaldes.
** Zentren der Rassen- und Artbildung.

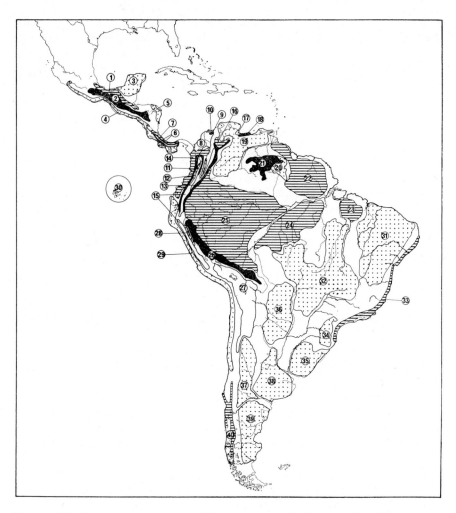

Fig. 101. The dispersal centres of terrestrial vertebrates in the Neotropical realm.

South America during the Pleistocene were more favourable than those of Africa (MARTIN 1966, MOREAU 1966). Moreover, because of the orographic subdivision of South America, Neotropical taxa were able to retreat north when the climate worsened or migrate downhill, as the montane forest centres did. The rate of extinction due to deterioration of climate may therefore have been lower than in the Holarctic region or Africa.

Furthermore, because the rain-forest biome is so heterogeneous (fig. 37), it is likely that competition would require close adaptation to niches without greatly increasing the extinction rate (MACARTHUR 1969, MACARTHUR & CONNELL 1970). In experiments on the effects of competition a species will only be eliminated if the environment remains homogeneous and if enough time is available (cf. GAUSE 1934, CROMBIE 1946, FRANK 1952, 1957). Competing species are able to coexist in environments which are not sufficiently homogeneous in space to cause extinction (CROMBIE 1946).

Variations in climate and competitive pressure did not therefore greatly raise the extinction rate in Neotropical rain forests. Competition only caused an important speciation factor to come into action, which was the process of adaptation to niches. A heterogeneous environment gives room for many co-existing species (MACARTHUR 1969).

But variations in climate, as I have shown, increased the mutual inter-penetration of the various biomes into each other, particularly of the non-forest and arboreal biomes. Continual displacements of these during the Pleistocene caused populations to become isolated. It is interesting to compare in this connection the interpretation of the distribution pattern of *Drosophila willistoni* in SPASSKY, RICHMOND et al (1971). The number of populations isolated was greater in the Neotropical realm than in other regions, as could be expected from the number of dispersal centres. Thus there are 40 dispersal centres in the Neotropical realm, but only 16 in the Nearctic region and only 22 in the Palaearctic region.

There are 3182 species of birds in South and Central America, on the basis of the works of EISENMANN (1955), MEYER DE SCHAUENSEE (1966), MONROE (1968) and HOWELL (1969). Of these only 1676 seem to be monocentric. All five of the Neotropical Pipid species are monocentric. There are 392 Leptodactylids in the South and Central America and of these at least 271 species are monocentric. And there are 74 species and subspecies of the genus *Bothrops* (HOGE 1966) and all these are monocentric except for *B. atrox*, *B. brazili*, *B. castelnaudi*, *B. nasutus*, *B. asper*, *B. schlegelli* and *B. bilineatus bilineatus*. Polycentric *Bothrops* species, moreover, are those for which very little comparative material exists.

It can thus be shown quantitatively that the richness in species of Neotropical environments correlates directly with the number of dispersal centres. It is therefore mistaken to try to solve the problem of why tropical biomes are so rich in species solely by invoking the principles of competition and of adaptation to niches (CODY 1969, KLOPFER & MACARTHUR 1961, MACARTHUR 1965, 1969). I do not wish to belittle the significance of these important evolutionary factors, but I should like to point out that the competition hypothesis of the authors mentioned pays too little attention to the theory of preadaptation (CUÉNOT 1932, KOSSWIG 1948, 1959, 1960, 1962, BOCK 1959, REMANE 1961, OSCHE 1962, MAYR 1967). The competition hypothesis does not clearly separate features which can be explained by competition from those acquired much earlier, when forms that now compete with each other were distributed allopatrically.

To sum up, strong displacements of the biochores have taken place during the Quaternary period and these displacements led to the formation of refuges. This is an essential reason for the richness of the tropical rain forest of Central and South America in species.

I have been able to show in what has preceded that the evolution of subspeci-fically distinct faunal elements can be used to work out the historical development of a dispersal centre. The ecological environments present in such a centre during a phase when it served as a refuge can often be judged from the ecological valency of the faunal elements now ascribed to it (cf. DE LATTIN 1957). This is the reason why **dispersal centres are also important for geographical studies.**

Thus there is a non-forest dispersal centre in the pampa of Argentina. This Pampa centre can be seen as the centre of origin of the subspecifically distinct populations within it, since these have an allopatric distribution. The existence of this dispersal centre therefore contradicts ELLENBERG's opinion (1962) that before the arrival of man the pampa was covered with forest.

This is confirmed by fossil evidence which shows that a fauna adapted to open landscapes lived in the Pleistocene where the Pampa centre now is (PATTERSON & PASCUAL 1968). Species that occur in the Pampa centre also extend into vegetational formations of scrub type (PARODI 1947, CABRERA 1957, FRENQUELLI 1946). But this also does not indicate that the pampa was originally tree-covered. On the contrary, the complete lack of forms that can be considered adapted to the arboreal biochore is much more important. It is true that man has lived in the pampa for 5000 years (WILLEY 1946, BORMIDA 1960) and has suppressed the continual growth of scrub by burning it back. Furthermore scrub fires are obviously capable of laying the pampa to waste (cf. HUDSON 1924). But man has not killed off the indicator species of open landscape.

PARODI (1930, 1940, 1942, 1947) invoked climate to explain the treelessness of the pampa. SCHMIEDER (1927), on the other hand, thought it was due to the scrub fires of the Indians. WALTER recognised (1966) that the pampa belonged to a semi-arid climatic zone. My zoogeographical results confirm and complement what WALTER (1966) and WEBER (1969) supposed, and leave no doubt that the east Argentinian pampa is fundamentally of grassland nature (MÜLLER 1972).

A non-forest dispersal centre in the high campos of the south-east Brazilian *Araucaria* forest i.e. the Parana centre, indicates that these high campos are also natural (cf. in this connection AUBRÉVILLE 1948, 1962, HUECK 1965, WEBER 1969).

A non-forest dispersal centre in the campo cerrado shows that this also represents a climax vegetation as WAIBEL (1921) first recognised. WALTER's opinion (1962), on the other hand, was that: 'the zonal vegetation of the campo cerrado would certainly be that of a rain forest which was destroyed by the native inhabitants even before the land was settled by whites'.* This opinion, however, completely contradicts the zoogeographical results here presented.

ELLENBERG (1958) supposed that the whole of the Puna centre had once been forested up to a height of 4500 m. (cf. also ELLENBERG 1966). But this view is contradicted by the position of the oreal Puna centre and the ecological adaptations of its faunal elements. These facts make it likely that, at least since the Older Peron (FAIRBRIDGE 1962), the Puna centre has existed essentially in its present form.

* 'die zonale Vegetation des Campo Cerrado sicher ein regengrüner Wald gewesen sei, die noch vor der Besiedlung durch Weiße von Eingeborenen zerstört wurde.'

V. SUMMARY

Compared with the Holarctic realm, the biogeographical study of South and Central America has only just begun, although numerous workers have concerned themselves with the plant and animal world of this area. The true reason for this slow start is the unparallelled richness in species of Neotropical environments. This explains why all previous proposals for the zoogeographical subdivision of the Neotropical realm have been based essentially on the various distributions of the ranges of families and genera. The zoogeographical regions, subregions and provinces so obtained, however, are put in question by the complexity of the ranges of species.

DE LATTIN (1957) developed the method of analysing dispersal centres. Up to the present this has only been applied in the Holarctic realm, but there its results have been confined by innumerable partial revisions. Analysis of dispersal centres begins from a comparative distributional viewpoint. It differs from the regional approach in being based solely and consistently on the geographical ranges of species and subspecies. The results got from analysing dispersal centres are not only significant for zoogeography, but also for the study of evolution and for geographical researches. In this way I have therefore undertaken to map out the dispersal centres in the Neotropical region for the terrestrial vertebrates, including amphibians, reptiles, birds and mammals.

A total of 40 dispersal centres can thus be shown to exist in the region that extends from the southern limit of DE LATTIN's Mexican centre (1957) south to Tierra del Fuego, but excluding the Antilles. The geographical position and faunal elements of these dispersal centres are described.

The affinities between centres can be analysed, based on the phylogeny of the species and subspecies which make up their faunal elements. Such an analysis shows that the dispersal centres can be divided into three groups. The faunal elements of group I are adapted to an unforested or, at most partly forested, biotope, generally below 1500 m. These centres belong to the non-forest biome. Their faunal elements are absent from the rain forest which, on the other hand, is inhabited by the faunal elements of the group II centres, belonging to the arboreal biome. The faunal elements of group III centres are strictly adapted to the oreal biome in the treeless high-mountain region.

The individual centres can be arranged according to affinity, into narrower subgroups within groups I, II and III. The montane forest centres form an especially well defined subgroup whose faunal elements are adapted to rain forest above 1400 or 1500 m. in altitude. The high-mountain oreal centres of Central America (Talamanca paramo centre) and northern South America (North Andean centre) are closest related to centres in the Nearctic region. The oreal Puna centre of the central part of the Andes has relationships both with the North Andean centre and with the Patagonian centre. The existence of a single Patagonian subregion cannot be confirmed.

For polytypic species with monocentric subspecies it can be shown that the dispersal centres represent centres where terrestrial vertebrates were pre-served during regressive phases of their environments, i.e. they have acted in the past as refuge areas. Subspecific differentiation of individual populations is a result of isolation from time to time in such refuges.

The position of dispersal centres has been decisively influenced by Quater-nary oscillations in climate and fluctuations in the vegetation.

There was a phase of expansion for open landscapes during the period 6000–2400 years ago which led to increased isolation of rain forests and of the faunas a-dapted to them. Campo species penetrated Amazonia for the last time during this phase. Their migration corridors were broken up by a renewed phase of forest ex-pansion beginning about 2400 years ago. The campo islands within the Amazonian rain forest, which have a specific campo fauna and flora, were connected by a continuous corridor of savanna during the post-glacial arid phase. They can be seen as relicts of this corridor.

Because of the strong warming in climate since younger Dryas times the biomes have shifted uphill. Consequently the faunas of the montane forests and of the oreal biome which about 11,000 years ago preferred a basimontane biotope, have suffered an increase in isolation. It is still not known what effect this up-ward shift of biomes had on the lowland rain-forest fauna, now found up to 1500 m. It is likely that much of the subspeciation of the montane forest fauna first started about 8000 years ago.

The large number of Neotropical distribution centres is correlated with the richness in species of the North and Central American biomes. The dispersal centres must also be seen as centres of origin for subspecies and species and their position has been greatly influenced by shifts of the biomes and climatic oscillations during the Quaternary. This supports the view that most species of the Neotropical forest fauna evolved in forest refuges during arid phases (HAFFER 1967, MÜLLER 1968).

The dispersal centres and their faunal elements can be used to indicate how the landscape evolved in the area in question. The non-forest centres shown to exist in the pampa of Argentina, in the campo cerrados and in the high campos of the Araucaria forests of Parana prove that these biotopes are ‚natural' and existed already in Postglacial times.

VI. BIBLIOGRAPHY

ABALOS, J. W., (1949): Cuales con los animales venenos de la Argentina? Tucuman.

ABALOS, J. W. and BAEZ, E. C., (1963): Variaciones del diseño en Bothrops neuwiedii meridionalis de Santiago del Estero. *Acta zool. Lilloana 19*: 479–486.

ABALOS, J. W., BAEZ, E. C. and NADER, R., (1964): Serpientes de Santiago del Estero (Republica Argentina). *Acta zool. Lilloana 20*: 211–283.

AB'SABER, A. N., (1962): Revisão dos conhecimentos sobre o horizonte subsuperficial de cascalhos inhumados do Brasil Oriental. Bol. Univ. Parana *2*.

AB'SABER, A. N., (1965): A evolução geomorfologica. In: A baixada Santista, São Paulo.

AB'SABER, A. N., (1970): Uma revisão do Quaternario Paulista: do presente para o passado. *Rev. Brasil. Geogr. 31* (4): 1–51.

ACOSTA Y LARA, E. F., (1950): Quiropteros del Uruguay. *Commun. Zool. Mus. Hist. Nat. Montevideo, 3* (58): 1–71.

AGUERREVERE, S. E., LOPEZ, V., DELGADO, C. and FREEMAN, C., (1939): Exploracion de la Gran Sabana. *Rev. de Fomento, Caracas, Venezuela, 3* (19): 501–729.

ALAYO, D. P., (1951): Especies herpetologicas halladas en Santiago de Cuba. *Bol. Hist. Nat. Soc. Habana, 2* (7): 106–110.

ALDRICH, J. and BOLE, B. P., (1937): The birds and mammals of the western slope of the Azuero peninsula (Republic of Panama). *Sci. Publ. Cleveland Mus. Nat. Hist. 7*: 1–196.

ALEMAN, G. C., (1952): Apuntes sobre reptiles y anfibios de la region Baruta-El Hatillo. *Mem. S. C. N. La Salle, Caracas 12* (31): 11–30.

ALEMAN, G. C., (1953): Contribución al Estudio de los Reptiles y Batracios de la Sierra Perijá. *Mem. Soc. Cienc. La Salle 13*: 499–531.

ALEXANDER, CH. S., (1961): The marine terraces of Aruba, Bonaire and Curaçao, Netherlands Antilles. *Annals of the Association of American Geographers 51* (1): 102–123.

ALLEN, G. F., (1902): The mammals of Margarita Island, Venezuela. *Proc. Biol. Soc. Washington 15*: 91–97.

ALMEIDA, F. F. M. DE, (1945): Geologia do sudoeste matogrossense. Dep. Nac. Prod. Min. *116*, Rio de Janeiro.

ALTMANN, P. L. and DITTMER, D. S., (1961): Blood and other body fluids. Fed. Amer. Soc. Exp. Biol. 1–191, Washington.

ALVAREZ, T., (1933): Observaciones biologicas sobre las aves del Uruguay. *An. Mus. Hist. Nat. Montevideo*, 2. Ser., *4* (1): 1–50.

ALVAREZ, T., (1934): Exterior y biologia de las aves uruguayas. Costumbres. Regimen alimentacio. *Su utilidad del punto de vista agricola. 3*: 1–64.

ALVIM, P. DE T., (1954): Teoria sôbre a formação dos Campos Cerrados. *Rev. Bras. Geogr. 16*.

AMADON, D., (1966): The superspecies Concept. *Syst. Zool. 15*: 245–249.

AMARAL, A. DO, (1927): Studies of neotropical Ophidia. *5*. Notes on Bothrops lansbergii and B. brachystoma. *Bull. Antiven. Inst. Amer. 1* (1).

AMARAL, A. DO, (1929): Studies of neotropical Ophidia. *8*. On the Bothrops lansbergii group. *Bull. Antiven. Inst. Amer. 3*: 19–27.

AMARAL, A. DO, (1935): Estudos sôbre ophidios neotropicos. *27*. Apontamentos sobre a fauna da Colombia. *Mem. Inst. But. 9*: 208–216.

AMARAL, A. DO, (1935): Estudos sôbre ophidios neotropicos. *28*. Novas especies de ophidios da Colombia. *Mem. Inst. But. 9*: 217–226.

ANDERSON, L. G., (1898): List of reptiles and batrachians collected by the Swedish expedition to Tierra del Fuego, 1895/96 under direction of Dr. OTTO NORDENSKJOLD. *Ofvers. Vet. Ak. Forsh.* 7: 457–662.

ANDRADE LIMA, D. DE, (1957): Estudos fitogeograficos de Pernambuco. Inst. Pesq. Agron. 2, Recife.

ANDRADE LIMA, D. DE, (1959): Viagem aos campos de Monte Alegre. *Bol. Inst. Agr. Norte, Belem.*

ANDREWS, E. W., (1937): Notes on snakes from the Yucatan Peninsula. *Zool. Ser. Field Mus. Nat. Hist.* 20: 355–359.

ANDRLE, R. F., (1967): The Horned Guan in Mexico and Guatemala. *The Condor 69* (2): 93–109.

ANT, H., (1964): Der boreoalpine Verbreitungstyp bei europäischen Landgastropoden. *Verhdl. Dtsch. Zool. Ges. Kiel.*

APLIN, O., (1894): On the birds of Uruguay. With an introduction and notes by P. L. SCLATER. *Ibis 6* (6): 149–215.

ARENS, K., FERRI, M. G. and COUTINHO, L. M., (1958): Papel do factor nutricional na economia d'agua de Plantas do Cerrado. *Rev. Biol. 1,* Rio de Janeiro.

ARISTEGUIETA, V., (1962): Una especie nueva de Senecio (Compositae) de la Gran Sabana. *Bol. Soc. Venez. Cien. Nat. 23:* 96–97.

ARKELL, A. J., (1964): 'Wanyanga'. Oxford University Press, London.

AUBRÉVILLE, A., (1948): La forêt de pin de Parana. *Bois et For. Trop.*

AUBRÉVILLE, A., (1962): Savanisation tropicale et glaciations quaternaires. *Adansonia 2:* 16–84.

AUER, V., (1933): Verschiebungen der Wald- und Steppengebiete Feuerlands in postglazialer Zeit. *Acta Geographica 5.*

AUER, V., (1960): The Quaternary history of Fuego-Patagonia. *Proc. Roy. Soc. London 152:* 507–516.

AUER, V., (1961): Die vulkanischen Schichten von Feuerland und Patagonien und das Zurückweichen der letzten Vergletscherung. *Schrft. Geogr. Inst. Kiel 20:* 277–289.

AUER, V., SALMI, M. and SALMINEN, K., (1955): Pollen and spore types of Fuego-Patagonia. *Ann. Acad. Sci. Fenn. 3* (43): 1–14.

AVERY, D. F. and TANNER, W. W., (1971): Evolution of the Iguanine Lizards (Sauria, Iguanidae) as determined by osteological and myological characters. Brigham Young Univ., *Sci. Bull. 12* (3): 1–79.

AVILA-PIRES, F. D. DE, (1969): Taxonomie und Zoogeographie des Genus 'Callithrix' ERXLEBEN, 1777 (Primates, Callithricidae). *Rev. bras. Biol.* 29: 49–64.

AZARA, F., (1847): Descripcion é historia del Paraguay y del Rio de la Plata. Madrid.

AZEVEDO, L. G. DE, (1962): Tipos de vegetação do sul de Minas e Campos da Mantiqueira. *Anais Acad. Bras. Cienc. 32,* Rio de Janeiro.

BADER, F., (1958): Die Verbreitung borealer und subantarktischer Holzgewächse in den Gebirgen des Tropengürtels. Diss. Bonn.

BAILEY, J. H., (1968): Spirorbinae (Polychaeta: Serpulidae) of the Galapagos Islands. *J. Zool. 155:* 161–184.

BAILEY, J. W., (1928): A revision of the lizards of the genus Ctenosaura. *Proc. U. S. Nat. Mus. 73:* 1–58.

BAKER, H. B., (1923): The Mollusca collected by the University of Michigan – Williamson Expedition in Venezuela. *Occ. Pap. Mus. Zool. Univ. Mich. 137:* 1–45.

BAKER, H. B., (1924): New Land Operculates from the Dutch Leeward Islands. *Nautilus 37:* 89–94.

BAKER, H. B., (1924): Land and Freshwater Molluscs of the Dutch Leeward Islands. *Occ. Pap. Mus. Zool. Michigan 152.*

BAKER, H. B., (1925): Isolation and Curaçao. *Nautilus 39:* 40–44.

BAKER, R. H., (1963): Geographical distribution of Terrestrial Mammals in Middle America. *Amer. Midl. Nat. 70* (1): 208–249.

BAKKER, E. M. VAN ZINDEREN, (1962): 'Palynology in Africa'. Seventh Report. University of the Orange Free State, Bloemfontein.

BAKKER, E. M. VAN ZINDEREN, (1962): Botanical evidence for quaternary climates in Africa. *Ann. Cape Prov. Mus. 2:* 16–31.

BAKKER, E. M. VAN ZINDEREN, (1963): Palaeobotanical studies. *S. Afr. J. Sci. 59*: 332-340.
BAKKER, E. M. VAN ZINDEREN, (1964): 'Palynology in Africa'. Eighth Report. University of the Orange Free State, Bloemfontein.
BAKKER, E. M. VAN ZINDEREN, (1964): A pollen diagram from equatorial Africa, Cherengani, Kenya. *Geologie Mijnb. 43*: 123-128.
BAKKER, E. M. VAN ZINDEREN, (1967): Upper Pleistocene stratigraphy and ecology on the basis of vegetation changes in sub-Saharan Africa. In Background to evolution in Africa. Chicago und London, Univ. Chicago Press.
BAKKER, I. P., (1954): Über den Einfluß von Klima, jüngerer Sedimentation und Boden-entwicklung auf die Savannen Nord-Surinams. *Erdkunde 8*, Bonn.
BANGS, O., (1898): A new murine opossum from Margarita Island. *Proc. Biol. Sc. Washington 12*: 95-96.
BANGS, O., (1909): Notes on some rare or not well-known Costa Rican birds. *Proc. Biol. Soc. Washington 22*: 29-38.
BARATTINI, L. P. and ESCALANTE, R., (1958): Catalogo de las aves uruguayas. *1*, Falconiformes. *Mus. Damaso A. Larrañaga* 1-102.
BARBOUR, TH., (1914): A contribution to the zoogeography of the West Indies, with special reference to amphibians and reptiles. *Mem. Mus. Comp. Zool. 44*: 209-346.
BARBOUR, TH., (1916): Amphibians and reptiles from Tobago. *Proc. Biol. Soc. Washington 29*: 221-224.
BARBOUR, TH., (1930): Antillean reptiles and amphibians. *Zoologica 11* (4), New York.
BARBOUR, TH., (1937): Third list of Antillean reptiles and amphibians. *Bull. Mus. Comp. Zool. 82* (2): 77-166.
BARBOUR, TH. and NOBLE, G. K., (1920): Some amphibians from northwestern Peru with a revision of the genera Phyllobates and Telmatobius. *Bull. Mus. Comp. Zool. 63*: 393-427.
BARBOUR, TH. and RAMSDEN, CH. T., (1919): The herpetology of Cuba. *Mem. Mus. Comp. Zool.* vol. *47*: 71-213.
BARBOUR, TH. and SHREVE, B., (1937): Novitates Cubanae. *Bull. Mus. Comp. Zool. 80*.
BARDEN, A., (1943): Food of the basilisk lizard in Panama. *Copeia 1943*: 118-121.
BARRERA, A., (1962): La peninsula de Yucatán como provincia biótica. *Rev. Soc. Mex. Hist. Nat. 23*.
BARRIO, A., (1964): Caracteres eto-ecológicos diferenciales entre Odontophrynus americanus (D. und B.) y O. occidentalis (BERG). (Anura, Leptodactylidae). *Physis 24*: 385-390.
BARRIO, A., (1964): Relaciones morfologicas, eto-ecologicas y zoogeograficas entre Physalae-mus henseli (PETERS) y P. fernandezae (MÜLLER). (Anura, Leptodactylidae). *Acta Zool. Lilloana 20*: 285-305.
BARRIO, A., (1965): El genero Physalaemus (Anura, Leptodactylidae) en la Argentina. *Physis 25*: 421-448.
BARROS, R., (1918): Notas sobre el sapito vaquero (Rhinoderma darwini). *Rev. Chil. Hist. Nat. 22*: 71-75.
BARROS, R. and CEI, J. M., (1962): The taxonomic position of three common Chilean frogs. *Herpetologica 18*: 195-203.
BARROWS, W. B., (1884): Birds of the lower Uruguay. *Auk 1*: 20-30, 109-113, 270-278, 313-319.
BARTHOLOMEW, G. A., (1966): A field study of temperature relations in the Galapagos marine iguana. *Copeia 1966* (2): 241-250.
BEARD, J. S., (1945): The Mora forests of Trinidad. *J. Ecol.*
BEARD, J. S., (1946): The Natural Vegetation of Trinidad. Oxford Forestry Memoirs 20.
BEARD, J. S., (1949): The Natural Vegetation of the Windward and Leeward Islands. Oxford Univ. Press.
BEARD, J. S., (1953): The Savanna Vegetation of Northern tropical America. Ecological Mono-graphs *23*, London.
BEARGIE, K. B. and MCCOY, C. J., (1964): Variation and Relationships of the Teiid Lizard Cnemi-dophorus angusticeps. *Copeia 1964* (3): 561-570.
BEAUMONT, J. DE, (1968): Zoogéographie des insectes de la Suisse. *Mitt. schweiz. ent. Ges. 41*: 323-329.

BEDERKE, E. and WUNDERLICH, H.-G., (1968): Geologiekarten. *Duden-Lexikon 3*: 513–536.

BEEBE, W., (1924): Galapagos: World's end. G. P. Putnam's Sons, New York.

BEEBE, W. and CRANE, J., (1948): Ecologia de Rancho Grande, Una selva nublade subtropical en el norte de Venezuela. *Bol. Soc. Ven. Cienc. Nat. 11* (73): 217–258.

BEEK, K. J. and BRAMAO, D. L., (1968): Nature and Geography of South American Soils. In: Biogeography and Ecology in South America *1*: 82–112.

BELDING, H., (1955): Geological development of the Colombian Andes. *Proc. Conf. Latin-Amer. Geol., Univ. Texas* 43–63.

BENNETT, A. G., (1926): A List of the birds of the Falkland Islands and Dependencies. Ibis *1*: 306–333.

BENNETT, A. G., (1931): Additional notes on the birds of the Falkland Islands and Dependencies. Ibis *2*: 12–13.

BENNETT, C. F., (1968): Human influences on the zoogeography of Panama. Ibero-Americana, Univ. Calif. Press, *51*.

BERG, C., (1898): Contribuciones al conocimiento de la fauna erpetologica Argentina. *An. Mus. Nac. Buenos Aires 4*: 1–35.

BERRY, E. W., (1925): Tertiary Flora of Trinidad. *Studies in Geology 6*: 71–161.

BERRY, E. W., (1925): A Pleistocene Flora from the Island of Trinidad. *Proc. U. S. Nat. Mus.* 2558.

BERRY, E. W., (1937): A Paleocene flora from Patagonia. Johns Hopkins University Studies in Geology, *12*, Baltimore.

BERTONI, A. DE WINKELRIED, (1914): Fauna Paraguaya. Catalogos sistematicos de los Vertebrados del Paraguay. *Descr. Fis. Econ. Paraguay 59* (1): 1–86.

BERTONI, A. DE WINKELRIED, (1928): El Crocodilurus o Jakarera en el Paraguay (Reptiles). *Rev. Soc. Cien. Paraguay 2*.

BERTONI, A. DE WINKELRIED, (1939): Catalogo sistematico de los Vertebrados del Paraguay. IV. Asuncion.

BESUCHET, C., (1968): Répartition des insectes en Suisse. Influence des glaciations. *Mitt. schweiz. ent. Ges. 41*: 337–340.

BEUCHER, F., (1963): Flores quaternaires au Sahara nord occidental, d'après l'analyse pollinique de sédiments prélevés à Hassi-Zguilma (Saoura). *C. r. hebd. Séanc. Acad. Sci. 256*, 2205–2208.

BIGARELLA, J. J., (1965): Subsidios para o estudo das variações de nivel oceanico no quaternario brasileiro. *An. Acad. Brasil. Ci. 37*: 263–278.

BIGARELLA, J. J., (1971): Variações climaticas no Quaternario Superior do Brasil e sua datação radiometrico pelo método do carbono 14. *Paleoclimas 1*: 1–22.

BIGARELLA, J. J. and ANDRADE, G. O. DE, (1965): Contribution to the study of the Brazilian Quaternary. International Studies of the Quaternary. *Geol. Soc. America 84*: 431–451.

BLAIR, W. F., (1963): Evolutionary relationships of North American Toads of the Genus Bufo. *Evolution 17*: 1–16.

BLAKE, E. R., (1953): Birds of Mexico. Univ. Chicago Press, Chicago.

BLAKE, E. R., (1955): A collection of Colombian game birds. *Fieldiana Zool. 37*: 9–23.

BLAKE, E. R., (1958): Birds of Volcain de Chiriqui. *Fieldiana Zool. 36*: 499–577.

BLAKE, E. R., (1961): New bird records from Surinam. *Ardea 49*: 178–183.

BLAKE, E. R., (1962): Birds of the Sierra Macarena, eastern Colombia. *Fieldiana Zool. 44*: 69–112.

BLAKE, E. R., (1963): The birds of southern Surinam. *Ardea 51*: 53–72.

BLYNDENSTEIN, J., (1967): Tropical savanna vegetation of the Llanos of Colombia. *Ecology 48* (1): 1–15.

BOCK, W. J., (1959): Preadaptation and multiple evolutionary pathways. *Evolution 13* (2): 194–211.

BOETTGER, O., (1885): Liste von Reptilien und Batrachiern aus Paraguay. *Z. f. Naturw. 4* (58): 213–248, 436–437.

BOETTGER, O., (1891): Reptilien und Batrachier aus Bolivia. *Zool. Anz. 14*: 343–347.

BOETTGER, O., (1893): Reptilien und Batrachier aus Venezuela. Ber. Senckenb. *Naturf. Ges.* 35–42.

BOETTGER, O., (1895): A contribution to the herpetological fauna of the island of Tobago. *J. Trinidad Field Natural.*, Port of Spain 2: 145–146.

BOKERMANN, W. C. A., (1964): Uma nova especie de Hyla da Serra do Mar em São Paulo (Amphibia, Salientia). *Rev. bras. Biol. 24*: 429–434.

BOKERMANN, W. C. A., (1964): Notes on tree frogs of the Hyla marmorata group with description of a new species. *Senck. biol. 45* (3/5): 243–254.

BOLDINGH, J., (1914): The flora of Curaçao, Aruba and Bonaire. Leiden.

BOND, G., (1963): Pleistocene environments in Southern Africa. In 'African Ecology and Human Evolution' (F. C. HOWELL and T. BOURLIÈRE, eds.), pp. 308–334. Viking Fund Publications in Anthropology, No. 36, New York.

BOND, J., (1960): Birds of the West Indies. London.

BOND, J. and MEYER DE SCHAUENSEE, R., (1942–43): The birds of Bolivia. *Proc. Acad. Nat. Sc. 94, 95* (1): 307–391, (2): 167–221.

BORMIDA, M., (1960): Investigaciones paleologicas en la region de Bolivar (Buenos Aires). An. Proc. Buenos Aires, Comis. Invest Ci. *1*.

BORRERO, J. I. and HERNANDEZ-CAMACHO, J., (1958): Apuntes sobre aves colombianas. *Caldasia 8* (37): 253–294.

BORRERO, J. I. and OLIVARES, A., (1955): Avifauna de la region de Soatá, departamento de Boyacá, Colombia. *Caldasia 7* (31): 51–81.

BORRERO, J. I., OLIVARES, A. and HERNANDEZ-CAMACHO, J., (1962): Notas sobre aves de Colombia. *Caldasia 8*: 585–601.

BOTT, R., (1958): Decapoden von den Galapagos-Inseln. *Senck. biol. 39* (3/4): 209–211.

BOULENGER, G. A., (1882): Catalogue of the Batrachia Salientia in the Collection of the British Museum ed. 2, London.

BOULENGER, G. A., (1894): List of Reptiles and Batrachians collected by Dr. BOHLS near Asuncion, Paraguay. *Ann. Mag. Nat. Hist. 6* (13): 342–348.

BOULENGER, G. A., (1898): A list of Reptiles, Batrachians and Fishes collected by Cap. Guido Boggiani in the Northern Chaco. *Ann. Mus. Civ. Stor. Nat. Genova 19*: 125–127.

BOULENGER, G. A., (1900): Descriptions of new Batrachians and Reptiles collected by Mr. P. O. SIMONS in Peru. *Ann. Mag. Nat. Hist. 7*: 181–186.

BOULENGER, G. A., (1901): Further descriptions of New Reptiles collected by Mr. P. O. SIMONS in Peru and Bolivia. *Ann. Mag. Nat. Hist. 7*: 546–549.

BOULENGER, G. A., (1902): Descriptions of New Batrachians and Reptiles from the Andes of Peru and Bolivia. *Ann. Mag. Nat. Hist. 10*: 394–402.

BOULENGER, G. A., (1903): On some Batrachians and Reptiles from Venezuela. *Ann. Mag. Nat. Hist. 11*: 481–484.

BOULENGER, G. A., (1905): Description of a new snake from Venezuela. *Ann. Mag. Nat. Hist. 7* (15).

BOWMAN, R. J., (1961): Morphological differentiation and adaptation in the Galapagos Finches. Univ. Calif. Press, Berkeley und Los Angeles.

BOWMAN, R. J., (1966): The Galapagos. Los Angeles, Berkeley.

BOYSON, V. F., (1924): The Falkland Islands with notes on the natural history by R. VALLENTIN. New York.

BRAME, A. H. and WAKE, D. B., (1963): The salamanders of South America. *Contr. Science 69*: 3–72.

BREYER, A., (1936): Lepidopteros de la Zona del Lago Nahuel Huapi, Territorio del Rio Negro. *Rev. soc. entom. arg. 8*: 61–63.

BREYER, A., (1939): Über die Argentinischen Pieriden. VII. Intern. Kongr. Entom., Berlin *1938*: 26–55.

BRICEÑO ROSSI, A. L., (1934): El problema del ofidismo en Venezuela. *Bol. Min. Salubr. Agric. Cria*, año II, *14*: 1079–1177.

BRODKORB, P., (1937): New or noteworthy birds from the Paraguayan chaco. *Occ. Pap. Mus. Zool. Univ. Michigan 345*: 1–2.

BRODKORB, P., (1954): A chachalaca from the Miocene of Florida. *Wilson Bull. 66*: 180–183.

BRODKORB, P., (1964): Catalogue of Fossil Birds. 2. *Bull. Flor. Stat. Mus. 8* (3): 195–335.

BRONGERSMA, L. D., (1940): Snakes from the Leeward Group, Venezuela and Eastern Colombia. Studies on the Fauna of Curaçao, *Aruba etc. 2*: 116–137.

BRONGERSMA, L. D., (1948): Frogs from the Leeward group, Venezuela and Eastern Colombia. In: HUMMELINCK, P. W., Studies on the Fauna of Curaçao, Aruba, Bonaire and the Venezuelan islands. The Hague *3*: 89–95.

BRONGERSMA, L. D., (1956): On some reptiles and amphibians from Trinidad and Tobago. *B. W. J. Proc. nederl. Akad. Wet. Amsterdam, ser.* C *59*: 165–188.

BRONGERSMA, L. D., (1966): Poisonous snakes of Surinam. *Mem. Inst. But. 33* (1): 73–79.

BROOKS, J. L., (1950): Speciation in ancient lakes. *Quart. Rev. Biol. 25*: 131–176.

BROSSET, A., (1963): La reproduction des oiseaux de mer des îles Galapagos en 1962. *Alauda 31* (2): 81–109.

BROWN, N. E., (1901): Report on two botanical collections made by Messrs. F. V. MC CONNELL and J. J. QUELCH at Mount Roraima in British Guiana. *Trans. Linnean Soc. London 6*: 1–107.

BRÜGGEN, J., (1929): Zur Glazialgeologie der chilenischen Anden. Geol. Rsch.

BRÜGGEN, J., (1934): Grundzüge der Geologie von Chile. Leipzig.

BRÜGGEN, J., (1948): La expansion del bosque en el sur de Chile en la epoca postglacial. *Rev. Univ. 28*, Santiago de Chile.

BRUNDIN, L., (1966): Transantarctic relationships and their significance, as evidenced by chironomid midges with a monography of the subfamilies Podonominae and Aphroteniinae and the austral Heptagyiae. K. Svenska Ventensk Akad. Handl. *11* (1).

BRUNDIN, L., (1972): Phylogenetics and Biogeography. *Syst. Zool. 21* (1): 69–79.

BRUNDIN, L., (1972): Evolution, Causal Biology, and Classification. *Zool. Script. 1*: 107–120.

BUDOWSKI, G., (1956): Tropical savannas, a sequence of forest felling and repeated burnings. Turrialba *6*, Turrialba.

BUISONJÉ, P. H. DE, (1964): Marine terraces and sub-aeric sediments on the Netherlands Leeward Islands Curaçao, Aruba and Bonaire, as indications of Quaternary changes in sea level and climate. *Koninkl. Nederl. Akad. Wetenschappen 67*: 60–79.

BUNTING, G. S., (1963): New species of Araceae from Chimanta Massif, Gran Sabana, Venezuela. *Bol. Venez. Cien. Nat. 25*: 29–33.

BÜRGL, H., (1961): Historia geologica de Colombia. *Rev. Acad. Col. Cien. Ex., Fis. y Nat. 11*: 137–191.

BURMEISTER, C. V., (1891): Relación de un viaje à la Gobernación de Chubut. *An. Mus. Nac. Buenos Aires 3*: 175–252.

BURMEISTER, H., (1861): Reise durch die La Plata-Staaten; ausgeführt in den Jahren 1857–60. 2 Bde., Halle.

BURT, CH. E., (1931): A study of the teiid lizards of the genus Cnemidophorus with special reference to their phylogenetic relationships. *Bull. U. S. Nat. Mus. 154*: 8–286.

BURT, CH. E., (1935): A new Lizard from the Dutch Leeward Islands (Cnemidophorus murinus ruthveni). *Occ. Pap. Mus. Zool. Michigan 324.*

BURT, CH. E. and BURT, M. D., (1933): A Preliminary check list of the Lizards of South America. Transact. *Acad. Science St. Louis 28* (1): 1–104.

BUTZER, K. W., (1957): Late Glacial and Postglacial Climatic Variation in the Near East. *Erdkunde 1*: 21–35.

BUTZER, K. W., (1961): Climatic change in arid regions since the Pliocene. In 'A History of Land Use in Arid Regions' (L. DUDLEY STAMP, ed.). UNESCO, Paris.

BUXBAUM, F., (1969): Der Entwicklungsweg der Kakteen in Südamerika. In: Biogeography and Ecology in South America, *2*: 583–623.

CABANIS, J. L., (1860–1862, 1869): Übersicht der im Berliner Museum befindlichen Vögel von Costa Rica. *J. Ornith. 8*: 321–336, 401–416, *9*: 1–11, 81–96, 241–256, *10*: 161–176, 321–336, *17*: 204–213.

CABRERA, A., (1957): Catalogo de los Mamiferos de America del Sur. *Rev. Mus. Argent. Cienc. Nat. Bern. Rivad. 4* (1): 1–307.

CABRERA, A. and YEPES, J., (1940): Mamiferos sudamericanos. Cia. Argent. Edit., Buenos Aires.

CABRERA, A. and YEPES, J., (1947): Zoogeografia. In 'Geografia de la Republica Argentina'. *Gaea 8*: 347–483.

CABRERA, A. L., (1958): La vegetación de la Puna Argentina. *Rev. Inv. Agric. 11* (4): 317–412.

CABRERA, A. L., (1958): Fitografia. In: La Argentina. *Suma de Geografia 3*: 101–207, Buenos Aires.

CALDENIUS, C. C., (1932): Las glaciaciones quaternarias en La Patagonia y Tierra del Fuego. *Geogr. Annal. 14*: 1–164.

CAMERON, A. W., (1958): Mammals of the islands in the Gulf of St. Lawrence. *Bull. Nat. Mus. Canada 154*: 1–165.

CAMPBELL, H. W. and HOWELL, T. R., (1965): Herpetological records from Nicaragua. *Herpetologica 21*: 130–140.

CAPOT-REY, R., (1961): Borkou et Ounianga. *Mém. Inst. Rech. sahar.*, No. 5.

CAPURRO, L. F., (1953): Distribution de Eupsophus taeniatus (GIRARD) en Chile. *Inv. Zool. Chil. 1, 10, 14/15.*

CAPURRO, L. F., (1954): El genero Telmatobius en Chile. *Rev. Chil. Hist. Nat. 54* (3): 31–40.

CAPURRO, L. F., (1957): Anfibios de la region de los lagos valdivianos. *Inv. Zool. Chil. 3*: 22–28.

CAPURRO, L. F., (1960): Eupsophus grayi de la isla Mocha. *Inv. Zool. Chil. 8.*

CARPENTER, CH., (1964): Comparative behavior of the Lava Lizards (Tropidurus) of the Galapagos Islands. *Amer. Zool. 4.*

CARPENTER, J. R., (1938): The Biome. *Amer. Midl. Natural. 21* (1).

CARR, A. F., (1950): Outline for a classification of animal habitats in Honduras. *Bull. Amer. Mus. Nat. Hist. 94*: 567–594.

CARVALHO, A. L. DE, (1954): A preliminary synopsis of the genera of the American microhylid frogs. *Occ. Pap. Mus. Zool., Univ. Mich. 555.*

CASAMIQUELA, R. M., (1958): Un anuro gigante del Mioceno de Patagonia. *Rev. As. Geol. Arg. 13* (3): 171–183).

CEI, J. M., (1950): Leptodactylus chaquensis n. sp. y el valor sistemático real de la especie lineana Leptodactylus ocellatus en la Argentina. *Acta Zool. Lilloana 9*: 395–423.

CEI, J. M., (1955): Chacoan Batrachians in Central Argentina. *Copeia 1955* (4): 291–293.

CEI, J. M., (1956): Nueva lista sistematica de los Batrachios de Argentina y breves notas sobre su biologia y ecologia. *Inv. Zool. Chil. 3*: 35–68.

CEI, J. M., (1957): Sobre la presencia de Pleurodema marmorata (D. u. B.) en territorio chileno. *Inv. Zool. Chil. 4.*

CEI, J. M., (1958): Polimorfismo y distribucion geografica en poblaciones chilenas de Pleurodema bibroni Tschudi. *Inv. Zool. Chil. 4*: 300–327.

CEI, J. M., (1959): La batracofauna chilena: muestra de procesos evolutivos. Acta I Congreso Sudamericano de Zool., La Plata.

CEI, J. M., (1960): A survey of the leptodactylid frogs, genus Eupsophus in Chile. *Breviora 118.*

CEI, J. M., (1960): Geographic variation of Bufo spinulosus in Chile. *Herpetologica 16*: 243–250.

CEI, J. M., (1961): Bufo arunco (Molina) y las formas chilenas de Bufo spinulosus Wiegmann. *Inv. Zool. Chil. 7*: 59–81.

CEI, J. M., (1962): Batracios de Chile. Ed. Univ. Chile, Santiago de Chile.

CEI, J. M., (1968): Distribution et spécialisation des batraciens sudaméricains. In: Biologie de l'Amérique australe, Paris.

CEI, J. M., (1969): The Patagonian Telmatobiid Fauna of the Volcanic Somuncura Plateau of Argentina. *J. Herpetology 3* (1–2): 1–18.

CEI, J. M. und BERTINI, F., (1961): Serum Proteins in Allopatric and Sympatric Populations of Leptodactylus ocellatus and L. chaquensi. *Copeia 1961* (3): 336–340.

CEI, J. M. and ERSPAMER, V., (1966): Biochemical Taxonomy of South American Amphibians by Means of Skin Amines and Polypeptides. *Copeia 1966* (1): 74–78.

CEI, J. M., ERSPAMER, V. and ROSEGHINI, M., (1967): Taxonomic and Evolutionary Significance of Biogenic Amines and Polypeptides occuring in Amphibian skin. I. Neotropical Leptodactylid frogs. *Syst. Zool. 16* (4): 328–342.

CHANEY, R. W., (1947): Tertiary centers and migration routes. *Ecol. Monogr. 17*: 141–148.

CHANG, J.-H., (1969): Some aspects of climatic fluctuations since the Pleistocene. *Geogr. Rev. 59* (4): 619–621.

CHAPIN, J. P., (1932): Birds of the Belgian Congo. *1. Amer. Mus. Nat. Hist. 65.*

213

CHAPMAN, F. M., (1917): The distribution of bird-life in Colombia: a contribution to a biological survey of South America. *Bull. Amer. Mus. Nat. Hist. 36*: 1–729.

CHAPMAN, F. M., (1921): The distribution of bird-life in the Urubamba valley of Peru. Smithsonian *Inst. U. S. Nat. Mus. Bull.* 117: 1–138.

CHAPMAN, F. M., (1923): The distribution of the motmots of the genus Momotus. *Bull. Amer. Mus. Nat. Hist. 48*: 27–59.

CHAPMAN, F. M., (1925): Remarks on the Life Zones of northeastern Venezuela with descriptions of new species of birds. *Amer. Mus. Novit. 191*: 1–15.

CHAPMAN, F. M., (1926): The distribution of bird-life in Ecuador, a contribution to the study of the origin of Andean bird-life. *Bull. Amer. Mus. Nat. Hist. 55*: 1–784.

CHAPMAN, F. M., (1929): Descriptions of new birds from Mt. Roraima. *Amer. Mus. Novit. 341*: 1–7.

CHAPMAN, F. M., (1931): The upper zonal bird-life of Mts. Roraima and Duida. *Bull. Amer. Mus. Nat. Hist. 63*: 1–135.

CHUBB, L. J., (1933): Geology of Galapagos, Cocos and Easter islands. *B. P. Bishop Mus. Bull. 110.*

CLARK, J. D., (1962): Carbon 14 chronology in Africa south of the Sahara. *Annls. Mis. r. Congo belge Ser 8*, vo. *40* (2): 303–314.

CLARK, J. D., (1963): Prehistoric Cultures of North-east Angola and their Significance in Tropical Africa. Companhia de Diamantes de Angola, Lisboa.

CLARK, J. D., (1964): The prehistoric origins of African culture. *J. Afr. Hist. 5*, 161–183.

CLARK, J. D. and FAGAN, B., (1965): Charcoals, sands and channel decorated pottery from Northern Rhodesia. *Am. Anthrop. 67*: 354–371.

COCHRAN, D. M., (1928): The herpetological Collections made in Haiti and its adjoining islands by WALTER J. EYERDAM. *Proc. Biol. Soc. Washington 41*: 53–59.

COCHRAN, D. M., (1941): The herpetology of Hispaniola. *Bull. U. S. Nat. Mus. 177*: 1–398.

COCHRAN, D. M., (1955): Frogs of southeastern Brazil. *U. S. Nat. Mus. 206*: 1–423.

COCHRAN, D. M., (1966): Taxonomy and Distribution of Arrow-Poison Frogs in Colombia. *Mem. Inst. But. 33* (1): 61–65.

CODY, M. L., (1969): Convergent characteristics in sympatric species: a possible relation to interspecific competition and aggression. *Condor 71*: 222–241.

COLE, L. J. and BARBOUR, TH., (1906): Vertebrates from Yucatan. Reptilia, Amphibia, Pisces. *Bull. Mus. Comp. Zool. 50*: 146–159.

COLE, M. M., (1960): Cerrado, Caatinga and Pantanal: the distribution and origin of the savanna vegetation of Brazil. *Geogr. Journ. 126*: 166–179.

COLLETTE, B., (1961): Correlation between ecology and morphology in anoline lizards from Havana, Cuba and southern Florida. *Bull. Mus. Comp. Zool. 125*: 137–162.

COOKE, H. B. S., (1964): The Pleistocene environment in southern Africa. In 'Ecological Studies in Southern Africa' (D. H. S. DAVIS, ed.), pp. 1–23. W. Junk, The Hague.

COOPE, G. R., (1969): The contribution that the Coleoptera of Glacial Britain could have made to the subsequent colonisation of Scandinavia. *Opusc. Ent. 34*: 95–108.

COOPER, J. E., (1958): Ecological notes on some Cuban lizards. *Herpetologica 14*: 53–54.

COPE, E. D., (1876): Report on the reptiles by Prof. James Orton from the Middle and Upper Amazon, and western Peru. *J. Acad. Nat. Science*, Philadelphia.

CORRINGTON, J. D., (1929): Herpetology of the Columbia, South Carolina, region. *Copeia*, no. 172, p. 58–83.

CORY, CH. B., (1909): The birds of the Leeward Islands, Caribbean Sea. *Field Mus. Nat. Hist. Publ. Ornith. 1* (5): 192–255.

CORY, CH. B., HELLMAYR, C. H. and CONOVER, B., (1918–1949): Catalogue of birds of the Americas. *Field Museum Natural History, 197* (13).

COUPER, R. A., (1960): Southern Hemisphere Mesozoic and Tertiary Podocarpaceae and Fagaceae and their palaeogeographic significance. *Proc. Roy. Soc. London 152*: 491–500.

COWAN, R. S., (1967): Rutaceae of the Guayana Highland. *Mem. New York Bot. Garden 14*: 1–14.

CROIZAT, L., (1958): Panbiogeography. Vol. I, IIa, IIb, Caracas.

214

CROMBIE, A. C., (1946): Further experiments on insect competition. *Proc. Roy. Soc. London 133*: 76–109.

CUATRECASAS, J., (1958): Aspectos de la vegetacion natural de Colombia. *Rev. Acad. Colombiana Cienc. Exact. Fis. Natur. 10* (40): 221–268.

CUELLO, J., (1959): Nuevos hallazgos de Picumnus nebulosus en el Uruguay. *Bol. Soc. Taguato 1* (2): 47–50.

CUELLO, J. and GERZENSTEIN, E., (1962): Las aves del Uruguay. Lista Sistematica, Distribucion y Notas. *Com. Zool. Mus. Hist. Nat. Montevideo 93* (6): 1–191.

CUÉNOT, L., (1932): La genèse des espèces animales. Paris.

CUNHA, O. R. DA, (1961): II. Lacertilios da Amazônia. *Bol. Mus. Goeldi 39*: 1–189.

CUNNINGHAM, R. O., (1871): Notes on the natural history of the Strait of Magellan and West Coast of Patagonia, made during the voyage of H. M. S. Nassau, in the years 1866–1869. *Edinburgh, 8*: 1–517.

CURIO, E., (1969): Funktionsweise und Stammesgeschichte des Flugfeinderkennens einiger Darwinfinken (Geospizinae). *Z. Tierpsychologie 26* (4): 394–487.

CURIO, E. and KRAMER, P., (1964): Vom Mangrovefinken (Cactospiza heliobates). *Z. Tierpsychologie 21*: 223–234.

DARBENE, R., (1926): Tres aves nuevas para la avifauna uruguaya. *El Hornero 3*: 422.

DAHL, E., (1960): The cold temperate zone in Chilean seas. *Proc. Roy. Soc. London 152*: 631–633.

DALL, W. H. and OSCHNER, W. H., (1928): Tertiary and Pleistocene Mollusca from the Galapagos Islands. *Proc. Calif. Acad. Sci. Ser. 4 17*: 89–139.

DANFORTH, ST., (1925): Porto Rican herpetological notes. *Copeia 147*: 76–79.

DANIEL, F., (1950): Liste der von Pater Cornelius Vogl in Maracay und Caracas gesammelten Schmetterlinge. *Bol. Ent. Venez. 8* (1/2): 21–42.

DANIEL, H., (1949): Las serpientes de Colombia. *Rev. Fac. Nac. Agron., Medellin 10* (36): 301–333.

DANSEREAU, P., (1947): Zonation e succession sur la restinga de Rio de Janeiro. *Rev. Canad. Biol. 6*, 448–477.

DARLINGTON, P. J., (1938): The origin of the fauna of the Greater Antilles, with discussion of dispersal of animals over water and through the air. *Quart. Rev. Biol. 13*: 274–300.

DARLINGTON, P. J., (1943): Carabidae of mountains and islands: Data on the evolution of isolated faunas, and on atrophy of wings. *Ecol. Monogr. 13*: 37–61.

DARLINGTON, P. J., (1957): Zoogeography. New York und London.

DARLINGTON, P. J., (1960): The zoogeography of the southern cold temperate zone. *Proc. Roy. Soc. London 152*: 659–668.

DARWIN, C., (1845): A journal of researches in natural history, kept during a voyage round the world. John Murray, London.

DARWIN, C., (1852): Journal of researches into the natural history and geology of the countries visited during the voyage of H.M.S. Beagle round the world. 2nd. Ed. London.

DARWIN, C., (1859): On the origin of species by means of natural selection, or the preservation of favoured races in the struggle for life. John Murray, London.

DAVIS, W. M., (1926): The Lesser Antilles. Amer. Geog. Soc, New York, 207 pp.

DAWSON, J. W., (1958): Interrelationships of the Australasian and South American floras. *Tuatara 7* (1): 1–6.

DEIGNAN, H. C., (1961): Type specimens of birds in the United States National Museum. *U.S. Nat. Mus. Bull. 221.*

DELANEY, P. J. V., (1963): Quaternary Geology History of the Coastal Plain of Rio Grande do Sul. Louisiana State Univ. Studies, Coastal Studies, Ser. 7.

DELANEY, P. J. V., (1966): Geology and geomorphology of the coastal plain of Rio Grande do Sul, Brazil and northern Uruguay. Baton Rouge.

DENBURGH, J. VAN, (1914): The gigantic land tortoises of the Galapagos Archipelago. *California Acad. Sci. Proc. 4* (2): 203–374.

DENBURGH, J. VAN and SLEVIN, J. R., (1913): Expedition of the California Academy of Sciences to the Galapagos Islands, 1905–1906. *Proc. Calif. Acad. Sciences 2* (1): 133–202.

DEVINCENZI, G. J., (1925–1928): Aves del Uruguay. *An. Mus. Hist. Nat. Montevideo, 2* (2): 129–200, 215–264, 339–407.

DEVINCENZI, G. J., (1935): Mamiferos del Uruguay. *An. Mus. Hist. Nat. Montevideo 4* (10): 1–96.

DICKEY, D. and VAN ROSSEM, A., (1938): The birds of El Salvador. *Publ. Field. Mus. Nat. Hist. zool. ser. 23.*

DIELS, L., (1934): Die Paramos der äquatorialen Hoch-Anden. Sitz. preuß. Akad. Wiss., *Physik.-math. Kl. 1934*: 57–68.

DOBSHANSKY, T., (1959): Evolution in the Tropics. *Scient. Amer. 38.*

DONAT, A., (1931): Über Pflanzenverbreitung und Vereisung in Patagonien. *Ber. Dt. Bot. Ges. 49.*

DONOSO-BARROS, R., (1961): Three new Lizards of the Genus Liolaemus from the highest Andes of Chile and Argentina. *Copeia 1961* (4): 387–391.

DONOSO-BARROS, R., (1961): The reptiles of the Lund University Chile Expedition. *Copeia* 1961 (4): 486–488.

DONOSO-BARROS, R., (1970): Catalogo herpetológico chileno. *Bol. Mus. Nac. de Hist. Nat. 31*: 49–124.

DORF, E., (1959): Climatic changes of the past and present. *Contrib. Mus. Paleo. Univ. Michigan 13* (8): 181–210.

DOWLING, H. G., (1960): A taxonomic study of the ratsnakes, genus Elaphe Fitzinger. VII. The triaspis section. *Zoologica. 45*: 53–80.

DUCKE, A. and BLACK, G. A., (1953): Phytogeographical notes on the Brazilian Amazon. *An. Acad. bras. Cienc. 25*: 1–46.

DUELLMAN, W. E., (1956): The frogs of the Hylid genus Phrynohyas Fitzinger, 1843. *Misc. Publ. Mus. Zool. Univ. Michigan 96.*

DUELLMAN, W. E., (1960): A Distributional study of the Amphibians of the Isthmus of Tehuantepec, Mexico. *Univ. Kansas Publ. 13* (2): 21–71.

DUELLMAN, W. E., (1963): A new species of tree frog, genus Phyllomedusa, from Costa Rica. Costa Rica: *Revista de Biol. Trop. 11*: 1–24.

DUELLMAN, W. E., (1963): Amphibians and reptiles of the rainforests of southern El Petén, Guatemala. *Univ. Kansas Publ., Mus. Nat. Hist. 15*: 205–249.

DUELLMAN, W. E., (1965): Amphibians and reptiles from the Yucatan Peninsula, Mexico. Univ. Kans. Pub., *Mus. Nat. Hist. 15* (12): 577–614.

DUELLMAN, W. E., (1966): The Central American Herpetofauna: An Ecological Perspective. *Copeia 1966* (4): 700–719.

DUGAND, A., (1952): Algunas aves del Rio Apoporis. Lozania (*Acta Zool. Colombiana) 4*: 1–12.

DUNN, E. R., (1926): The frogs of Jamaica. *Proc. Boston Soc. nat. Hist. 38*: 111–130.

DUNN, E. R., (1928): Notes on Bothrops lansbergii and Bothrops ophryomegas. *Bull. Antiven. Inst. Amer. 2* (2): 29–30.

DUNN, E. R., (1931): The herpetological fauna of the Americas. *Copeia 1931* (3): 106–119.

DUNN, E. R., (1932): The colubrid snakes of the Greater Antilles. *Copeia* 1932, pp. 89–92.

DUNN, E. R., (1940): Some aspects of herpetology in lower Central America. *Trans. N. Y. Acad. Sci. 2*: 156–158.

DUNN, E. R., (1944): A revision of the Colombian snakes of the genera Leimadophis, Lygophis, Liophis, Rhadinaea and Pliocercus, with a note on Colombian Coniophanes. *Caldasia 2* (10): 479–495.

DUNN, E. R., (1944): A review of the Colombian snakes of the families Typhlopidae and Leptotyphlopidae. *Caldesia 3* (11): 47–55.

DUNN, E. R., (1944): Herpetology of the Bogota Area. Rev. Acad. Cienc. *Colombia 6* (21): 68–81.

DUNN, E. R., (1949): Relative abundance of some Panamanian snakes. *Ecology 30* (1): 39–57.

DURHAM, J. W., ARELLANO, A. R. V. and PECK, J. H., (1952): No Cenozoic Tehuantepec seaways. *Bull. Geol. Soc. Amer. 63*: 1245.

DURHAM, J. W., ARELLANO, A. R. V. and PECK, J. H., (1955): Evidence for no Cenozoic Isthmus of Tehuantepec seaway. *Bull. Geol. Soc. Amer. 66*: 977–992.

216

ECHTERNACHT, A. C., (1968): Distributional and ecological notes on some reptiles from northern Honduras. *Herpetologica 24*: 151–158.

ECHTERNACHT, A. C., (1970): Taxonomic and ecological notes on some Middle and South American lizards of the genus Ameiva (Teiidae). *Breviore 354*: 1–9.

ECHTERNACHT, A. C., (1971): Middle American lizards of the Genus Ameiva (Teiidae) with emphasis on geographic variation. *Univ. Kansas, Miscell. Publ. 55*: 1–86.

ECKEL, E. B., (1952): General geology of Paraguay. Report US operation mission. Asuncion.

EIBL-EIBESFELDT, J., (1962): Neue Unterarten der Meerechse, Amblyrhynchus cristatus, nebst weiteren Angaben zur Biologie der Art. *Senck. biol. 43* (3): 177–199.

EIBL-EIBESFELDT, J., (1966): Beobachtungen über das innerartliche Kampfverhalten der Kielschwanzleguane (Tropidurus) des Galapagos-Archipels. *Z. Tierpsychologie 6*: 672–676.

EIGENMANN, C. H., (1905): The fresh-water fishes of Patagonia and an examination of the Archiplate-Archhelenis Theory. Rept. Princeton Univ. Exped. to Patagonia (1896–1899) *3* (2).

EISENMANN, E., (1955): The species of Middle American birds. Trans. Linnean Soc. New York, 7.

EISENMANN, E., (1962): Notes on some Neotropical vireos in Panama. *Condor 64*: 505–508.

EISENTRAUT, M., (1931): Biologische Studien im bolivianischen Chaco. I. Die Reise mit kurzem Überblick über Landschaft, Bevölkerung und Tierwelt. – S.-B. *Ges. naturf. Freunde Berlin 4/7*: 167–192.

EISENTRAUT, M., (1935): Biologische Studien im bolivianischen Chaco. VI. Beitrag zur Biologie der Vogelfauna. – *Mitt. Zool. Mus. Berlin 20*: 307–443.

EISENTRAUT, M., (1968): Die tiergeographische Bedeutung des Oku-Gebirges im Bamenda-Banso-Hochland (Westkamerun). *Bonn. Zool. Beitr. 19*: 170–175.

EISENTRAUT, M., (1970): Eiszeitklima und heutige Tierverbreitung im tropischen Westafrika. *Umschau 3*: 70–75.

EISENTRAUT, M., (1973): Die Wirbeltiere von Fernando Poo und Westkamerun. *Bonner Zool. Monogr.*

ELLENBERG, H., (1958): Wald oder Steppe? Die natürliche Pflanzendecke der Anden Perus. Umschau in Wissenschaft u. Technik *21*.

ELLENBERG, H., (1959): Typen tropischer Urwälder in Peru. Schweiz. Z. f. Forstwesen *3*.

ELLENBERG, H., (1962): Wald in der Pampa Argentiniens? Festschr. F. FIRBAS. Veröff. Geob. Inst. Rübel *37*.

ELLENBERG, H., (1966): Leben und Kampf an den Baumgrenzen der Erde. *Verhdl. Vb. Dtsch. Biol. 3*: 35–41.

ELTON, C., (1930): Animal ecology and evolution. Oxford.

EMDEN, F. VAN, (1935): Die Carabiden der Deutschen Chaco-Expedition (Col.). Rev. Entomologia, Rio de Janeiro.

EMILIANI, C., (1955): Pleistocene Temperatures. *J. Geol. 63* (6): 538–578.

EMSLEY, M. G., (1964): Speciation in Heliconius (Lep. Nymphalidae): Morphology and Geographic Distribution. *Zoologica 49*: 191–254.

EMSLEY, M. G., (1966): The status of the snake Erythrolamprus ocellatus Peters. *Copeia 1966*: 128–129.

EMSLEY, M. G., (1966): The mimetic significance of Erythrolamprus aesculapii ocellatus Peters from Tobago. *Evolution 20*: 663–664.

ERIKSEN, W., (1972): Störungen des Ökosystems patagonischer Steppen- und Waldregionen unter dem Einfluss von Klima und Mensch. Biogeographica *1*: 57–73.

ESCALANTE, R., (1959): Some records of oceanic birds in Uruguay. *Condor 61* (2): 158–159.

ESCALANTE, R., (1960): Occurence of the Osprey in Uruguay. *Condor 62* (2): 138.

ESCALANTE, R., (1961): Occurence of the Cassin Race of the Peregrine Falcon in Uruguay. *Condor 63* (2): 180.

ESPINAL, T., LUIS, S. and MONTENEGRO, E. M., (1963): Formaciones vegetales de Colombia. Memoria explicativa sobre el mapa ecologico. Inst. Geogr. Agustin Codazzi, Dept. Agr., Bogota.

ESPINOZA, R., (1932): Ökologische Studien über Kordillerenpflanzen. Dissertation, Jena.

ESTES, R., (1961): Miocene lizards from Colombia, South America. *Breviora 123*: 1–11.

ESTES, R. and WASSERSUG, R., (1963): A Miocene toad from Colombia, South America. *Breviora, Mus. Comp. Zool. 193*: 1–13.

ETHERIDGE, R., (1965): Fossil lizards from the Dominican Republic. *Quart. Jour. Florida Acad. Sci. 28* (1): 83–105.

EVANS, H. E., (1947): Notes on Panamanian reptiles and amphibians. *Copeia 1947*: 166–170.

EWAN, J., (1950): Ferns of Pico Bolivar and the sources of the Venezuelan flora. *Amer. Flm. Jour. 40*: 109–116.

FAIRBRIDGE, R. W., (1958): Dating the latest movements of the Quaternary sea level. *Trans. N. Y. Acad. Sci. 20*: 471–482.

FAIRBRIDGE, R. W., (1960): The changing level of the sea. *Sci. American 5*: 70–79.

FAIRBRIDGE, R. W., (1961): Eustatic changes in sea level. *Physics and Chemistry of the Earth 4*: 99–185.

FAIRBRIDGE, R. W., (1961): Convergence of evidence on climatic change and ice age. *Am. N. York Acad. Sci. 95*: 542–579.

FAIRBRIDGE, R. W., (1962): World Sea-Level and climatic changes. *Quaternaria 6*: 111–134.

FAIRBRIDGE, R. W., (1967): Climatic variations. In: Encyclopedia of atmospheric Sciences and Astrogeology, 205–211. New York.

FERNANDEZ, K., (1927): Sobre la biologia y reproduccion de Batraciós argentinos. II. *Bol. Ac. Nac. Cienc. Cordoba 29*: 271–328.

FERNANDEZ YEPEZ, A., (1946): Le Avifauna Venezolana y su Distribucion en Zonas Altitudinales. *Acta Venezolana 1* (3): 1–8.

FERRI, M. G., (1954): Water-Balance of the Caatinga, a semi-arid type of vegetation of northern Brazil. 7 Cong. Intern. Bot. Paris.

FERRI, M. G., (1955): Contribuição ao conhecimento da ecologia do Cerrado e da Caatinga. São Paulo.

FERRI, M. G., (1960): Contribution to the knowledge of the Ecology of the 'Rio Negro Caatinga' (Amazon). Bul. Research Council of Israel.

FERRI, M. G., (1963): Historica dos trabalhos botanicos sobre o Cerrado. Simp. Cerrado, São Paulo.

FERRI, M. G. and COUTINHO, L. M., (1960): Contribuição ao conhecimento da ecologia do Cerrado. Fac. Fil. Cienc. e Letras, São Paulo.

FIRBAS, F., (1964): Die glazialen Refugien der europäischen Gehölze (ohne Osteuropa). Report 6th Int. Congr. on Quaternary Warsaw, 375–382, Lódz.

FITTKAU, E. J., (1969): The fauna of South America. In: Biogeography and Ecology in South America 2: 624–658.

FLORIN, R., (1940): The Tertiary fossil conifers of South Chile and their phytogeographical significance. *K. Svensk. Vet. Akad. Hdl. 19* (2): 1–107.

FORSTER, W., (1958): Die tiergeographischen Verhältnisse Boliviens. Proc. tenth internat. Congress of Entomol. *1*: 843–846.

FOUQUETTE, M. J., (1968): Observations on the Natural History of Microteiid Lizards from the Venezuelan Andes. *Copeia 1968* (4): 881–884.

FOWLER, H., (1913): Amphibians and reptiles from Ecuador, Venezuela and Yucatan. *Proc. Acad. Nat. Sci. Philadelphia 65*: 153–176.

FOWLER, H., (1918): Some amphibians and reptiles from Porto Rico and the Virgin Islands. *Pap. Dept. Marine Biol. Carnegie Inst. 12*: 1–15.

FOWLER, H., (1944): Results of the Fifth George Vanderbilt Expedition (1941) (Bahamas, Caribbean Sea, Panama, Galápagos Archipelago and Mexican Pacific Islands). *Acad. Nat. Sci. Phil., Monogr. 6*: 57–529.

FRANK, P., (1952): A laboratory study of intraspecies and interspecies competition in Daphnia pulicaria (FORBES) and Simocephalus vetulus (O. F. MÜLLER). *Physiol. Zool. 25*: 173–204.

FRANK, P., (1957): Coactions in laboratory populations of two species of Daphnia. *Ecology 38*: 510–519.

FRANKLIN, H. J., (1912): The Bombidae of the New World. *Trans. amer. ent. Soc., 39* (2): 73–100.

FRANTZIUS, A. VON, (1869): Ueber die geographische Verbreitung der Vögel Costaricas und deren Lebensweise. *J. Ornith. 17*: 195–204, 289–318, 361–379.

218

FRAY, CH. and EWING, M., (1963): Pleistocene Sedimentation and Fauna of the Argentine .Shelf *Proc. Acad. Nat. Sci. Phil. 115*: 113–126.

FREIBERG, M. A., (1942): Enumeracion sistematica y distribucion geografica de los batracios argentinos. *Physis 19*: 219–240.

FRENQUELLI, J., (1946): Las grandes unidades fisicas del territorio argentino. In: Geografia de la Republica Argentina *3*: 5–360, Buenos Aires.

FRENZEL, B., (1968): Grundzüge der pleistozänen Vegetationsgeschichte Nord-Eurasiens. Franz Steiner Verl., Wiesbaden.

FRIEDMANN, H., GRISCOM, L. and MOORE, R. T., (1950): Distributional check-list of the birds of Mexico. Part I. Pacific Coast Avif. (Cooper Orn. Soc.) *29*: 1–202.

FRITTS, T. H., (1969): The Systematics of the parthenogenetic Lizards of the Cnemidophorus cozumela Complex. *Copeia 1969* (3): 519–535.

FRY, C. H., (1970): Ecological distribution of birds in northeastern Mato Grosso State, Brazil. *An. Acad. brasil. Ciênc. 42*: 275–318.

GAIGE, H. T., (1936): Some reptiles and amphibians from Yucatan and Campeche, Mexico. *Carnegie Inst. Washington Publ. 457*: 289–304.

GALLARDO, J. M., (1960): Estudio zoogeografico del genero Leiosaurus (Reptile, Sauria). *Physis*, 113–118.

GALLARDO, J. M., (1961): Three new toads from South America: Bufo manicorensis, Bufo spinulosus altiperuvianus and Bufo quechua. *Breviora, Mus. Comp. Zool. 141*.

GANS, C. and ALEXANDER, A., (1962): Studies on amphisbaenids (Amphisbaenia, Reptilia). 2. On the amphisbaenids of the Antilles. *Bull. Mus. Comp. Zool.*, Harvard Univ. *128* (3): 65–158.

GANSSER, A., (1954): The Guiana Shield (South America). *Eclogae Geol. Helvetiae. 47*: 77–117.

GAUSE, G. F., (1934): The struggle for existence. Williams und Wilkins, Baltimore.

GEIJSKES, D. C., (1934): Notes on the Odonate Fauna of the Dutch West Indian Islands Aruba, Curaçao and Bonaire. *Int. Rev. Hydrobiol. 31*: 287–311.

GENTILLI, J., (1949): Foundations of Australian bird geography. *Emu 49*: 85–129.

GERY, J., (1969): The Fresh-Water Fishes of South-America. In: Biogeography and Ecology in South America, *2*: 828–848.

GILLIARD, E. TH., (1941): The birds of Mt. Auyan-tepui, Venezuela. *Bull. Amer. Mus. Nat. Hist. 77*: 439–508.

GIRARD, CH., (1854): Abstract of a report to Lieut. James M. Gillis upon the reptiles collected during the U.S.N. Astron. Expedition to Chili. Proc. Acad. Natur. Science, Philadelphia.

GLEASON, H. A., (1929): A collection of plants from Mt. Duida. *J. New York Bot. Garden* 30 (355).

GLEASON, H. A., (1929): The Tate collection from Mt. Roraima and vicinity. *Bull. Torrey. Bot. Club 56*: 391–408.

GLOYD, H. K., (1940): The rattlesnakes, genera Sistrurus and Crotalus. *Chicago Acad. Sci. Special Publ. 4*: 7–266.

GODLEY, E. J., (1960): The Botany of southern Chile in relation to New Zealand and the Sub-antarctic. *Proc. Royal Soc.*, London B, *152*.

GOEBEL, K., (1891): Die Vegetation der venezolanischen Paramos. In: Pflanzenbiologische Schilderungen *2*: 1–50, Marburg.

GOETSCH, W., (1930): Observaciones y experimentos con animales chilenos. An. Univ. Chile, 977–1018.

GOETSCH, W., (1930): Expediciones informativas por el pais para el estudio de la fauna chilena. An. Univ. de Chile.

GOETSCH, W., (1933): Verbreitungsverhältnisse chilenischer Eidechsen, Ameisen und Planarien. *Forsch. u. Fortschr. 9*: 66–67.

GOETSCH, W., (1933): Verbreitung und Biologie der Landplanarien Chiles. *Zool. Jb.*, Abt. *Syst., Ökol und Geogr. Tiere 64*: 245–288.

GOETSCH, W., (1933): Die chilenischen Termiten. *Zool. Jb., Abt. Syst. Ökol. und Geogr. Tiere, 64*: 227–244.

219

GOETSCH, W., (1933): Übersicht über die biologischen Exkursionen. Fauna Chilensis II. *Zool. Jb. Syst. 64*: (2) 153–164.

GOETSCH, W. and HELLMICH, W., (1932): Variabilität bei chilenischen Eidechsen und Fröschen. *Ztschr. Indukt. Abst. u. Vererb.-lehre 52*: 67–72.

GOETSCH, W. and HELLMICH, W., (1933): Chilenische Landschaften und ihre Charaktertiere. *Peterm. Geogr. Mitt. 9/10*: 239–242.

GOODALL, J. D., JOHNSON, A. W. and PHILIPPI, R. A., (1946–1964): Las aves de Chile. Santiago de Chile.

GOODLAND, R., (1966): On the savanna vegetation of Calabozo, Venezuela and Rupununi, British Guiana. *Bol. Soc. Venezolana Cienc. Natur. 26* (110): 341–359.

GOODLAND, R., (1971): A Physiognomic Analysis of the 'Cerrado' vegetation of Central Brasil. *J. Ecol.* 411–418.

GOODWIN, G., (1946): Mammals of Costa Rica. *Bull. Amer. Mus. Nat. Hist. 87*: 271–474.

GOOSEN, D., (1964): Geomorfologia de los Llanos orientales. *Rev. Acad. Col. Cien. Ex., Fis. y Nat. 12*: 129–139.

GORHAM, ST. W., (1966): Liste der rezenten Amphibien und Reptilien. Ascaphidae, Leiopelmatidae, Pipidae, Discoglossidae, Pelobatidae, Leptodactylidae, Rhinophrynidae. *Das Tierreich 85*: 1–222.

GORMAN, G. C. and ATKINS, L., (1969): The Zoogeography of Lesser Antillean 'Anolis' lizards. An analysis based upon Chromosomes. *Bull. Mus. Comp. Zool. 138*.

GORMAN, J., (1968): Breeding of an andean toad in Hot Springs. *Copeia 1*: 167–170.

GOULD, ST. J., (1969): Character variations in two land snails from the Dutch Leeward Islands: Geography, Environment and Evolution. *Syst. Zool. 18*: 185–200.

GRANDIDIER, G. and NEVEU-LEMAIRE, M., (1908): Observations relatives à quelques tatous rares ou inconnus habitant la 'puna' Argentine et Bolivienne. *Bull. Mus. Hist. Nat. 14*: 4–7.

GRANT, CH., (1932): Herpetology of Tortola; notes on Anegada and Virgin Gorda, British Virgin Islands. *Jour. Dept. Agric. Puerto Rico 16*: 339–346.

GRIFFITHS, I., (1959): The phylogeny of Sminthillus limbatus and the status of the Brachycephalidae (Amphibia, Salientia). *Proc. Zool. Soc. London 132*: 457–487.

GRISCOM, L., (1932): The distribution of bird-life in Guatemala. *Bull. Amer. Mus. Nat. Hist. 64*: 1–439.

GRISCOM, L., (1935): The ornithology of the Republic of Panama. *Bull. Mus. Comp. Zool. 78*: 261–382.

GROSS, F. J., (1961): Zur Geschichte und Verbreitung der euro-asiatischen Satyriden (Lepidoptera). *Verhdl. Dtsch. Zool. Ges. Bonn* (1960).

GROSS, F. J., (1962): Zur Evolution euro-asiatischer Lepidopteren. *Verhdl. Dtsch. Zool. Ges. Saarbrücken* (1961).

GUICHENOT, (1848): Reptiles. In GAY, Historia fisica y politica de Chile. *Zool. 2*: 5–136.

HAFFER, J., (1959): Notas sobre las aves de la region de Uraba. *Lozania (Acta Zool. Colombiana) 12*: 1–49.

HAFFER, J., (1967): On the Dispersal of Highland Birds in tropical South and Central America. *El Hornero 10* (4): 436–438.

HAFFER, J., (1967): On birds from the northern Choco region, N.W.-Colombia. *Veröff. Zool. Staatssamml.*, München. *11*: 123–149.

HAFFER, J., (1967): Some allopatric species pairs of birds in Northwestern Colombia. *Auk 84* (3): 343–365.

HAFFER, J., (1967): Speciation in Colombian Forest birds West of The Andes. *Amer. Mus. Novitates 2294*: 1–57.

HAFFER, J., (1967): Zoogeographical notes on the 'Nonforest' Lowland Bird Faunas of Northwestern South America. *El Hornero 10* (4): 315–333.

HAFFER, J., (1969): Speciation in Amazonian Forest Birds. *Science 165*: 131–137.

HAFFER, J., (1970): Entstehung und Ausbreitung nord-andiner Bergvögel. *Zool. Jb. Syst. 97*: 301–337.

HAFFER, J., (1970): Art-Entstehung bei einigen Waldvögeln Amazoniens. *J. Ornith. 111* (3/4): 285–331.

HAFFER, J., (1970): Geologic-climatic history and zoogeographic signifiance of the Uraba Region in Northwestern Colombia. *Caldasia 10* (50): 603–636.

HAFFER, J., (1971): Nachtrag zur Verbreitung von Pipra fasciicauda und Pipra iris in Brasilien. *J. Orn. 112* (4): 460–461.

HAFFER, J. and BORRERO, J. J., (1965): On birds from northern Colombia. *Rev. Biol. Tropical, Costa Rica 13*: 29–53.

HAMILTON, T. H. and RUBINOFF, I., (1963): Isolation, endemism, and multiplication of species in the Darwin finches. *Evolution 17*: 388–404.

HAMILTON, T. H. and RUBINOFF, I., (1964): On models predicting abundance of species and endemics for the Darwin Finches in the Galapagos Archipelago. *Evolution 18*: 339–342.

HARDY, L. M. and COLE, CH. J., (1967): The colubrid snake Tantilla armillata COPE in Nicaragua. *J. Ariz. Acad. Science 4* (3): 194–196.

HARDY, L. M. and MCDIARMID, R. W., (1969): The Amphibians and Reptiles of Sinoloa, Mexico. *Univ. Kansas Publ. 18* (3): 39–252.

HARTERT, E., (1893): On the birds of the islands of Aruba, Curaçao and Bonaire. Ibis *19*: 289–338, London.

HARTERT, E. and VENTURI, S., (1909): Notes sur les oiseaux de la République Argentine. – Nov. Zool. *16*.

HARTWEG, N., (1934): Description of a new Kinosternid from Yucatán. *Occ. Papers Mus. Zool. Univ. Michigan 277*: 1–2.

HASTENRATH, ST. L., (1967): Observations on the snow line in the Peruvian Andes. *J. Glaciology 6*: 541–550.

HAUMAN, L., (1913): La forêt valdivienne et ses limites. *Rec. Inst. Bot. Leo Errera 9*: 346–408.

HAUMAN, L., (1918): La végétation des hautes cordillères. *An. Soc. bien. Arg. 86*: 121–128, 225–348.

HAUMAN, L., (1931): Esquisse phytogéographique de l'Argentine subtropicale et de ses relations avec la Géobotanique sudaméricaine. *Bull. Soc. Bot. Belg. 64*: 20–80.

HAVERSCHMIDT, F., (1968): Birds of Surinam. Oliver und Boyd, Edinburgh und London.

HELLEBREKERS, W. PH. J., (1945): Further notes on the Penard Zoological collection from Surinam. *Zoöl. Mededelingen 25*: 93–100.

HELLMAYR, C. E., (1912): Zoologische Ergebnisse einer Reise in das Mündungsgebiet des Amazonas. II. Vögel. *Abh. Königl. Bayer. Akad. Wissensch. Math.-physik. Kl. 26* (2): 1–142.

HELLMICH, W., (1933): Biologische Exkursionen in den Hochanden Chiles. *Der Biologe 2*: 133–136.

HELLMICH, W., (1933): Die biogeographischen Grundlagen Chiles. *Zool. Jahrb.*, Abt. Syst., Oekol., Geogr. Tiere (Fauna Chil. II), 165–226.

HELLMICH, W., (1934): Die Eidechsen Chiles, insbesondere die Gattung Liolaemus. *Abhdl. Bayer. Akad. Wissensch. 24*: 1–140.

HELLMICH, W., (1938): Beiträge zur Kenntnis der Herpetofauna Chiles. XII. Die Eidechsen des Volcán Villarica. *Zool. Anz. 124*: 237–249.

HELLMICH, W., (1940): Contribucion al Conocimiento de la Fauna Venezolana. *Bol. Soc. Ven. Cienc. Nat. 6* (46): 318–327.

HELLMICH, W., (1950): Über die Liolaemus-Arten Patagoniens. *Ark. Zool. 1* (22): 345–353.

HELLMICH, W., (1953): Contribucion al conocimiento de los ofidios de Venezuela. 1. Spilotes pullatus pullatus (L.). *Acta Biol. Venezuelica 1* (8): 141–145.

HELLMICH, W., (1953): Contribucion al conocimiento de los ofidios de Venezuela. 2. Sobre la subespecie venezolana de Coluber (Masticophis) mentovarius (D. u. B.). *Acta Biol. Venezuelica 1* (8): 146–154.

HELLMICH, W., (1960): Die Sauria des Gran Chaco und seiner Randgebiete. *Abhdl. Bayer. Akad. Wiss., Math.-Naturw. Kl. 101*: 1–131.

HELLMICH, W., (1962): Bemerkungen zur individuellen Variabilität von Liolaemus multiformis (COPE). *Opuscula Zoologica 67*: 1–10.

HENNIG, W., (1960): Die Dipteren-Fauna von Neuseeland als systematisches und tiergeographisches Problem. *Beitr. Ent. 10*.

HERSHKOVITZ, PH., (1958): A geographical Classification of neotropical mammals. *Fieldiana: Zool. 36* (6): 581–620.

HERSHKOVITZ, PH., (1966): Mice, land bridges and Latin America faunal interchange. In: R. L. WENZEL and V. J. TIPTON, Parasites of Panama, 725–747. Field Museum Natur. Hist., Chicago.

HERSHKOVITZ, PH., (1969): The evolution of mammals on southern continents. VI. The recent mammals of the neotropical region: A Zoogeographic and ecological review. *Quart. Rev. Biol. 44*: 1–70.

HERZOG, TH., (1923): Die Pflanzenwelt der bolivianischen Anden. Vegetation d. Erde *15*, Leipzig.

HETTNER, A., (1881): Das Klima von Chile und Westpatagonien. Dissertation, Bonn.

HOBSON, E. S., (1969): Remarks on aquatic habits of the Galapagos marine Iguana, including submergence times, cleaning symbiosis, and the Shark threat. Copeia *1969* (2): 401–402.

HOGE, A. R., (1962): Serpentes da Fundação 'Surinam Museum'. *Mem. Inst. But. 30*: 51–64.

HOGE, A. R., (1966): Preliminary account on neotropical Crotalinae (Serpentes, Viperidae). *Mem. Inst. Butantan 32*: 109–184.

HOLDGATE, M. W., (1960): The Royal Society Expedition to southern Chile. *Proc. Royal Soc., London B 152.*

HOLDRIDGE, L. R., (1953): La vegetación de Costa Rica. In 'Atlas estadistico de Costa Rica'. San José.

HOLDRIDGE, L. R., (1964): Life zone ecology. San José, Costa Rica: Tropical Science Center.

HOLLICK, A., (1924): A Review of the Fossil Flora of the West Indies. *Bull. N. Y. Bot. Gard. 12*: 259–323.

HONEGGER, R. E., (1972): Die Reptilien-Bestände auf den Galapagos-Inseln 1972. *Nat. u. Mus. 102* (12): 437–454.

HOOIYER, D. A., (1959): Fossil rodents from Curação and Bonaire. Studies on the Fauna of Curação and other Caribbean Islands, *9* (35): 1–27.

HOOKER, J. D., (1847): On the vegetation of the Galapagos Archipelago. *Transact. Linn. Soc.*

HOVANITZ, W., (1945): Comparisons of some Andean butterfly faunas. *Caldasia 3*: 301–36.

HOVANITZ, W., (1945): Distribution of Colias in the Equatorial Andes. *Caldasia 3*: 283–300.

HOVANITZ, W., (1958): Distribution of Butterflies in the New World. Zoogeography, Amer. Assoc. Advanc. Science.

HOWELL, TH. R., (1957): Birds of a second-growth rain forest area of Nicaragua. *Condor 59*: 73–111.

HOWELL, TH. R., (1969): Avian distribution in Central America. *Auk 86*: 293–326.

HOY, G., (1971): Über Brutbiologie und Eier einiger Vögel aus Nordwest-Argentinien II. *J. Orn. 112* (2): 158–163.

HUBACH, E., (1958): Estratigrafia de la Sabana de Bogotá y alrededores. *Bol. Geologico 5* (2): 93–112.

HUBER, J., (1896): Über die Vegetation des Küstengebiets von Brasilisch-Guyana. *Bol. Mus. Paraense, 1*: 381–401.

HUBER, J., (1900): Sur les campos de l'Amazone inférieur et leur origine. Extrait du Compte-rendu du Congr. internat. de botanique à l'Expos. Univ., 387–400, Paris.

HUBER, J., (1902): Zur Entstehungsgeschichte der brasilianischen Campos. *Peterm. Mitt. 48*: 92–95.

HUDSON, H., (1920): Birds of La Plata. London.

HUDSON, H., (1924): Far away and long ago. A history of my early life. *1*, New York.

HUECK, K., (1957): Sobre a origem dos Campos Cerrados no Brasil e algunas novas observações no seu limite meridional. *Rev. Bras. Geogr. 14*, Rio de Janeiro.

HUECK, K., (1965): Der Araukarien- und Podocarpus Wald und das Kamp-Problem von Campos do Jordão in der Serra de Mantiqueira. Compt. rendus *18* Congr. Int. Geogr. Rio de Janeiro.

HUECK, K., (1966): Die Wälder Südamerikas. Verl. Fischer, Stuttgart.

HUGOT, H. J., QUÉZEL, P. and MARTINEZ, O., (1962): Documents scientifiques. Mission Berliet Ténéré-Tchad, Paris.

HUMBOLDT, A. VON, (1859): Reise in die Aequinoctialgegenden des neuen Continents. Stuttgart.

HUMMELINCK, P. W., (1933): Reisebericht. Zoologische Ergebnisse einer Reise nach Bonaire, Curação und Aruba im Jahre 1930. *Zool. Jb. Syst. 64*: 289–326.

222

HUMMELINCK, P. W., (1938): Notes on the Cactaceae of Curaçao, Aruba, Bonaire and North Venezuela. *Rec. Trav. Bot. Neerl. 35*: 29–55.

HUMMELINCK, P. W., (1939): Apuntaciones sobre las Aguas Superficiales del Estado Nueva Esparta y Dependencias Federales. *Bol. Soc. Venez. Cienc. Nat. 5*: 173–178.

HUMPHREY, PH. S., BRIDGE, D., REYNOLDS, P. W. and PETERSON, R. T., (1970):Birds of Isla Grande (Tierra del Fuego). Smithsonian Manual. Publ.: Off:, Lawrence, Kansas.

HURT, W. R., (1964): Recent radiocarbon dates for Central and Southern Brazil. *Amer. Antiq. 30*: 25–33.

HUTCHINSON, G. E., (1965): The ecological theater and the evolutionary play. Yale Univ. Press, New Haven.

HUTCHINSON, G. E., PATRICK, R. and DEEVEY, E. S., (1956): Sediments of Lake Patzcuaro, Michoacan, Mexico. *Bull. Geol. Soc. Amer. 67*: 1491–1504.

IHERING, H. VON, (1907): Distribuição de Campos e matas no Brasil. Rev. Mus. Paul., São Paulo.

ILLIES, J., (1960): Phylogenie und Verbreitungsgeschichte der Ordnung Plecoptera. Verh. Dtsch. Zool. Ges., Bonn.

ILLIES, J., (1961): Versuch einer allgemeinen biozönotischen Gliederung der Fließgewässer. *Int. Rev. Ges. Hydrobiol. Hydrogr. 46*: 205–213.

ILLIES, J., (1965): Die Wegnersche Kontinentalverschiebungstheorie im Lichte der modernen Biogeographie. *Naturwissenschaften 52* (18): 505–511.

ILLIES, J., (1966): Neue Theorien über die Kontinentaldrift. *Zool. Anz. 177*: 46–50.

ILLIES, J., (1969): Biogeography and Ecology of Neotropical Freshwater Insects, especially those from Running Waters. In: Biogeography and Ecology in South America *2*: 685–708.

ILLIES, J., (1969): Revision der Plecopterenfamilie Austroperlidae. *Ent. Tidskrift Arg. 90* (1/2): 19–51.

JOHANSEN, H., (1969): Nordamerikanische Zugvögel in der Südhälfte Südamerikas. *Bonner Zool. Beitr. 1/3*: 182–190.

JOHNSON, A. W., (1965): The birds of Chile. *1*: 1–398. Santiago de Chile.

JOHNSTON, J. R., (1909): Flora of the Islands of Margarita and Coche, Venezuela. *Proc. Boston Soc. Nat. Hist. 34*: 163–312.

JOHNSTON, R. F., (1969): Taxonomy of house sparrows and their allies in the Mediterranean basin. *Condor 71*: 129–139.

JOHNSTON, R. F., (1969): Character variation and adaption in European sparrows. *Syst. Zool. 18*: 206–231.

JOHNSTON, R. F., (1970): Intralocality character dimorphism in house sparrows. Symp. on Passer Research, S.C. Kendeigh, Ed.

JOHNSTON, R. F. and SELANDER, R. K., (1964): House sparrows: rapid evolution of races in North America. *Science 144*: 548–550.

JOHNSTON, R. F. and SELANDER, R. K., (1971): Evolution in the House Sparrow. II. Adaptive Differentiation in North American Populations. *Evolution 25* (1): 1–28.

JUNGE, G. C. A. and MEES, G. F., (1961): The Avifauna of Trinidad and Tobago. E. J. Brill, Leiden.

KANTER, H., (1936): Der Gran Chaco und seine Randgebiete. *Abh. Gebiet Auslandskde. 43*: 1–376, Hamburg.

KÄSTLE, W., (1965): Zur Ethologie des Anden-Anolis Phenacosaurus richteri. *Z. Tierpsychologie 22* (7): 751–769.

KATZER, F., (1898): Eine Forschungsreise nach der Insel Marajó. *Globus 73*.

KATZER, F., (1902): Zur Frage der Entstehung der brasilianischen Campos. *Peterm. Mitt. 48*: 190–191.

KEAST, A., (1959): Vertebrate speciation in Australia: some comparisons between birds, marsupials and reptiles. In: C. W. Leeper (ed.), Evolution of Living Organisms, 380–407. Symp. Royal Soc. Victoria, Melbourne University Press.

KEAST, A., (1961): Bird speciation on the Australian continent. *Bull. Mus. Com. Zool. Harvard Coll. 123*: 305–495.

KEAST, A., (1968): Evolution of mammals on southern continents. IV, Australian mammals: Zoogeography and Evolution. *Quaterly Rev. Biol. 43* (4): 373–408.

223

KENNY, J. S., (1969): The Amphibia of Trinidad. Stud. Fauna Curaçao. Den Haag, 29 (108): 1–78.

KEPKA, O., (1969): Zur Tiergeographie der Trombiculidae im Mittelmeerraum. *Verhl. Dtsch. Zool. Ges. 32*: 526–535.

KINSEY, E. C., (1942): The breeding of Galapagos finches in California. *Aviculture Mag.*, ser. 5, *8*: 125–137.

KLAUBER, L. M., (1945): The geckos of the genus Coleonyx with descriptions of new subspecies. *Trans. San Diego Soc. Nat. Hist. 10*: 133–216.

KLAUBER, L. M., (1956): Rattlesnakes. Their habits, life histories, and influence on mankind. Berkeley and Los Angeles.

KLOPFER, P. and MACARTHUR, R. H., (1961): On the causes of tropical species diversity: Niche overlap. *Amer. Naturalist 95*: 223–226.

KNOCHE, W., (1927): Karten der Januar- und Juli-Bewölkung in Chile. *Z. Ges. Erdk.* 220–224, Berlin.

KNOCHE, W., (1929): Jahres-, Januar- und Juli-Niederschlagskarten der Republik Chile. *Z. Ges. Erdk.* 208–216, Berlin.

KNOX, G. A., (1960): Littoral ecology and biogeography of the southern oceans. *Proc. Roy. Soc. London 152*: 577–624.

KOEGEL, L., (1922): Zur Frage der Urwaldentwicklung in Amazonien. *Geogr. Z. 28*: 187–190.

KOEPCKE, H.-W., (1957): Über die Wälder an der Westseite der peruanischen Anden und ihre tiergeographischen Beziehungen. *Verhdl. Dtsch. Zool. Ges. 21*: 108–119.

KOEPCKE, H.-W., (1961): Synökologische Studien an der Westseite der peruanischen Anden. *Bonner Geogr. Abhdl. 29*: 1–320.

KOEPCKE, H.-W. and KOEPCKE, M., (1963): Las aves silvestres del Peru. Ministero Agricultura, Lima.

KOEPCKE, M., (1967): Zur Kenntnis einiger Furnariiden (Aves) der Küste und des westlichen Andenabhangs Perus. *Beitr. Neotr. Fauna 4* (3): 150–173.

KOEPCKE, M., (1970): The birds of the department of Lima, Peru. Liv. Publ. Comp.

KOFORD, C. B., (1957): The vicuna and the puna. *Ecol. Monogr. 27*: 153–219.

KOOPMAN, K. F., (1958): A fossil Vampire Bat from Cuba. *Breviora 90*: 1–4.

KOSLOWSKY, J., (1895): Batracios y Reptiles de Rioja y Catamarca recognidos durante los meses de febrero a mayo de 1895. *Rev. Mus. La Plata 6*.

KOSLOWSKY, J., (1896): Sobre algunos reptiles de Patagonia y otras regiones argentinas. *Rev. Museo La Plata 7*: 447–457.

KOSLOWSKY, J., (1898): Enumeración sistematica y distribucion geografica de los reptiles argentinos. *Rev. Museo La Plata 8*: 161–200.

KOSSWIG, C., (1936): Die Evolution von Anpassungsmerkmalen bei Höhlentieren. *Zool. Anz. 112*: 148–155.

KOSSWIG, C., (1942): Die Faunengeschichte des Mittel- und Schwarzen Meeres. *C. Soc. Turqu. Sci. Phys. Nat. 9*.

KOSSWIG, C., (1948): Homologe und analoge Gene, parallele Evolution und Konvergenz. *Comm. Fac. Sci. Univ. Ankara 1*: 33–39.

KOSSWIG, C., (1948): Genetische Beiträge zur Präadaptationstheorie. *Rev. Fac. Sci. Univ. Istanbul 13*: 176–209.

KOSSWIG, C., (1959): Phylogenetische Trends, genetisch betrachtet. *Zool. Anz. 162* (7/8): 208–221.

KOSSWIG, C., (1960): Zur Phylogenese sogenannter Anpassungsmerkmale bei Höhlentieren. *Int. Rev. Ges. Hydrobiol. 45*: 493–512.

KOSSWIG, C., (1960): Bemerkungen zum Phänomen der intralakustrischen Speziation. *Zool. Beitr. 5* (2): 497–512.

KOSSWIG, C., (1962): Über präadaptive Mechanismen in der Evolution vom Gesichtspunkt der Genetik. *Zool. Anz. 169* (1/2): 4–14.

KOSSWIG, C. and VILLWOCK, W., (1964): Das Problem der intralakustrischen Speziation im Titicaca- und im Lanaosee. Verh. Zool. Ges. Kiel.

KRAUS, O., (1954): Myriapoden aus Peru II. *Senck. biol. 35* (1/2): 17–55.

KRAUS, O., (1955): Taxonomische und tiergeographische Studien an Myriapoden und Araneen aus Zentralamerika. Dissertation, Frankfurt.

KRAUS, O., (1960): Zur Zoogeographie von Zentral-Amerika (Studien an Myriapoden und Arachniden). Verhdl. XI. Intern. Kongr. Entomol. *1*: 516–518, Wien.

KRAUS, O., (1964): Tiergeographische Betrachtungen zur Frage einer einstigen Landverbindung über den Südatlantik. Nat. und Mus. *94*.

KRIEG, H., (1924): Biologische Reisestudien in Südamerika. II. Rhinoderma und Calyptocephalus. *Z. Morph. Ökol. Tiere 3* (1): 150–168.

KRIEG, H., (1927): Die tiergeographischen Probleme des Gran Chaco. *Zool. Anz. 74*: 271–283.

KRIEG, H., (1931): Wissenschaftliche Ergebnisse der deutschen Gran Chaco-Expedition. Geographische Übersicht und illustrierter Routenbericht. Stuttgart.

KRIEG, H., (1936): Tiergeographische Wirkungen der jährlichen Überschwemmungen im Stromgebiet des Paraguay und Parana. *Mitt. d. Isis*, Doppelheft, München.

KRIEG, H., (1936): Der Gran Chaco als tiergeographisches Problem. *Compt. Rend. du XII. Congrès Internat. Zool.* Lisbonne.

KRIEG, H., (1939): Als Zoologe in Steppen und Urwäldern Patagoniens. München.

KUHLMANN, E., (1958): Vegetação campestre do Planalto Meridional do Brasil. Inst. Bras. Géo. et Est.

KÜHNELT, W., (1969): Zur Ökologie der Schneerandfauna. *Verhdl. Dtsch. Zool. Ges. 32*: 707–721, Innsbruck.

KUSCHEL, G., (1960): Terrestrial zoology in southern Chile. *Proc. Roy. Soc., London, B, 152*: 540–550.

KUSCHEL, G., (1969): Biogeography and Ecology of South American Coleoptera. In: Biogeography and Ecology in South America 2: 709–722.

LACK, D., (1947): Darwin's Finches. Cambridge Univ. Press.

LANE, J., (1943): The geographic distribution of Sabethini (Dipth. Culicidae). *Rev. Entom. 14* (3): 409–429.

LANGE, M. DE, FERRI, B. and FERRI, M. G., (1964): Contribuição ao estudo da anatomia do plantas do Cerrado. *Bol. Fac. Cienc. Letras, 234*, São Paulo.

LANGGUTH, A., (1969): Die südamerikanischen Canidae unter besonderer Berücksichtigung des Mähnenwolfes Chrysocyon brachyurus. Illiger. *Z. wiss. Zool. 179* (1/2): 1–188.

LANJOUW, J., (1936): Studies of the vegetation of the Surinam Savannas and Swamps. *Med. Bot. Mus. 33*, Utrecht.

LANNING, E. P., (1965): Early Man in Peru. *Scient. Americ. 213* (4): 68–76.

LANNING, E. P., (1967): Peru before the Incas. Prentice-Hall, Englewood, New York.

LANYON, W., (1967): Revision and probable Evolution of the Myiarchus Flycatchers of the West Indies. *Bull. Amer. Mus. Nat. Hist. 136* (6): 331–370.

LATASTE, F., (1891): Etudes sur la faune chilienne. *Anal. Soc. scient. Chil. 1*: 3–40.

LATTIN, G. DE, (1939): Über die Evolution von Höhlentiercharakteren. 1. Sitz.—Ber. d. Ges. Naturf.—Freunde Berlin, 11–41.

LATTIN, G. DE, (1952): Zur Evolution der westpaläarktischen Lepidopterenfauna. Decheniana *105/106*.

LATTIN, G. DE, (1957): Die Ausbreitungszentren der holarktischen Landtierwelt. *Verhdl. Dtsch. Zool. Ges. Hamburg.*

LATTIN, G. DE, (1959): Postglaziale Disjunktionen und Rassenbildung bei europäischen Lepidopteren. *Verhdl. Dtsch. Zool. Ges. Frankfurt.*

LATTIN, G. DE, (1967): Grundriss der Zoogeographie. Jena.

LAUBMANN, A., (1930): Wissenschaftliche Ergebnisse der deutschen Gran Chaco-Expedition, Vögel. Stuttgart.

LAUBMANN, A., (1939/40): Die Vögel von Paraguay. *1*: 1–246, 2: 1–228, Stuttgart.

LAUER, W., (1952): Humide und aride Jahreszeiten in Afrika und Südamerika und ihre Beziehungen zu den Vegetationsgürteln. In: Studien zur Klima- und Vegetationskunde der Tropen. *Bonner Geogr. Abhdl. 9*.

LAUER, W., (1954): Las formas de la vegetacion de El Salvador. *Comun. Inst. Trop. Invest. Cient. 3* (1): 41–45.

LAUER, W., (1956): Vegetation, Landnutzung und Agrarpotential in El Salvador. (Zentralamerika). *Schr. Geogr. Inst. Univ. Kiel 16* (1).

LAUER, W., (1959): Klimatische und pflanzengeographische Grundzüge Zentralamerikas. *Erdkunde 13*.

LAUER, W., (1960): Probleme der Vegetationsgliederung auf der mittelamerikanischen Landbrücke. *Dt. Geogr. Berlin 32*: 123–132, Franz Steiner Verlag, Wiesbaden.

LAZELL, J. D., (1964): The Anoles (Sauria, Iguanidae) of the Guadeloupéen Archipelago. *Bull. Mus. Comp. Zool. 131* (11): 359–401.

LAZELL, J. D., (1964): The Lesser Antillean Representatives of Bothrops and Constrictor. *Bull. Mus. Comp. Zool.*, Harvard Univ. *132* (3): 245–273.

LAZELL, J. D. and WILLIAMS, E. E., (1962): The anoles of the eastern Caribbean (Sauria, Iguanidae). Parts IV–VI. *Bull. Mus. Comp. Zool.*, Harvard Coll., *127* (9): 451–478.

LEGLER, J. M., (1963): Tortoises (Geochelone carbonaria) in Panama: Distribution and Variation. *Amer. Midl. Nat. 69*: 490–503.

LELEUP, N. and LELEUP, J., (1968, 1970): Mission zoologique belge aux Iles Galapagos et en Ecuador, 1964 et 1965. *1, 2*, Brüssel.

LEOPOLD, A. ST., (1959): Wildlife of Mexico. Univ. Calif. Press, Berkeley und Los Angeles.

LINDROTH, C. H., (1970): The theory of glacial refugia in Scandinavia. Comments on present opinions. *Notul. Entomol. 44*: 178–192.

LINELL, M. L., (1899): On the coleopterous insects of the Galapagos Islands. *U. S. Nat. Mus. Proc. 21*: 249–268.

LLIBOUTRY, L., (1956): Nieves y glaciares de Chile. Ed. Univ. Chile, Santiago de Chile, 1–471.

LORIE, J., (1887): Fossile Mollusken von Curaçao, Aruba und der Kueste von Venezuela. *Samml. Geol. Reichsmus. Leiden 1*: 111–149.

LÖTSCHERT, W., (1955): La vegetación de El Salvador. Communicaciones del Instituto tropical de investigaciones cientif. *4*: 65–79.

LOWE, P. R., (1907): On the birds of Margarita island, Venezuela. Ibis *1* (9): 547–570, London.

LOWERY, G. H. and DALQUEST, W. W., (1951): Birds from the state of Veracruz, Mexico. *Univ. Kansas Publ. Mus. Nat. Hist. 3*: 531–649.

LUCAS, F. A., (1900): Characters and relations of Gallinuloides, a fossil gallinaceous bird from the Green River shales of Wyoming. *Bull. Mus. Comp. Zool. 36*: 79–84.

LUDWIG, G., (1966): Probleme im Paläzoikum des Amazones- und des Maranhao-Beckens in erdölgeologischer Sicht. Erdöl. u. Kohle, Erdgas, *Petrochemie 19*: 798–807.

LUDWIG, G., (1968): Die geologische Entwicklung des Marajó-Beckens in Nordbrasilien. *Geol. Jb. 86*: 845–878.

LUDWIG, W., (1954): Die Selektionstheorie. In HEBERER, Die Evolution der Organismen, 662–712. G. Fischer, Stuttgart.

LUTZ, A., (1927): Notas sobre batrachios de Venezuela e da Ilha de Trinidad. *Mem. Inst. Osw. Cruz 20* (1): 35–50.

LUTZ, B., (1952): New frogs from Itatiaia Mountain, Brazil. *Copeia 1952* (1): 27–28.

LUTZ, B., (1963): New species of Hyla from Southeastern Brazil. *Copeia 1963* (3): 561–562.

LUTZ, B. and BOKERMANN, W. C., (1963): A new Tree Frog from Santa Catarina, Brazil. *Copeia 1963* (3): 558–561.

LYNCH, J. D., (1968): The status of the Nominal Genera Basanitia and Phrynanodus from Brazil (Amphibia: Leptodactylidae). *Copeia 1968* (4): 875–876.

LYNCH, J. D. and FUGLER, C. M., (1965): A survey of the frogs of Honduras. *J. Ohio Herp. Soc. 5* (1): 5–18.

MAACK, R., (1958): Vorläufige Ergebnisse einer Forschungsreise durch Südafrika zum Problem der tangentialen Krustenverschiebungen der Erde. *Die Erde 89*: 284–305.

MAACK, R., (1962): Neue Forschungen in Paraguay und am Rio Parana. Die Flußgebiete Monday und Acaray. *Die Erde 93*: 4–48.

MACARTHUR, R. H., (1965): Patterns of species diversity. *Biol. Rev. 40*: 510–533.

MACARTHUR, R. H., (1969): Patterns of communities in the tropics. *Biol. J. Linn. Soc. 1*: 19–30.

MACARTHUR, R. H. and CONNELL, J. H., (1970): Biologie der Populationen. BLV, München, Basel, Wien.

226

MACARTHUR, R. H. and WILSON, E. O., (1963): An equilibrium theory of insular zoogeography. *Evolution 17*: 373–387.

MACARTHUR, R. H. and WILSON, E. O., (1967): The theory of island biogeography. Princeton Univ. Press, 1–203.

MACHATSCHEK, F., (1955): Das Relief der Erde 2: 1–594, Berlin.

MAGUIRE, B., (1960): Floristic exploration and research in South American Guyana Highland. *Year Book, Amer. Phil. Soc.* 317–321.

MAGUIRE, B., (1965): The botany of the Guyana Highland. VI. *Mem. New York Bot. Garden 12* (3): 1–285.

MALME, G. O., (1924): Beiträge zur Kenntnis der Cerrado-Bäume von Mato Grosso. *Ark. f. Botanik 18*, Stockholm.

MANN, G., (1968): Die Ökosysteme Südamerikas. In: Biogeography and Ecology in South America, *1*: 171–229.

MARCUZZI, G., (1950): Ofidios existentes en las colecciones de los museos de Caracas (Venezuela). Noved. Cient. *La Salle 3*: 1–20.

MARSHALL, W. H., (1940): A survey of mammals of the islands in Great Salt Lake. Utah. *J. Mammal. 21*: 144–159.

MARTIN, P. S., (1958): Pleistocene Ecology and Biogeography of North America. In: *Zoogeography*, Washington, 375–420.

MARTIN, P. S., (1966): Africa and Pleistocene overkill. *Nature 212*: 339–342.

MARTIN, P. S. and HARRELL, B. E., (1957): The Pleistocene history of temperate biotas in Mexico and eastern Unites States. *Ecology 38*: 468–480.

MARTIN, S. W., (1965): Glacial Lakes in the Bolivian Andes. *Geogr. J. 131*: 519–526.

MASLIN, T. P., (1963): Notes on a collection of herpetozoa from the Yucatan Peninsula of Mexico. *Univ. Colorado Studies, Biol. 9*: 1–20.

MATURANA, H. R., (1962): A study of the species of the genus Basiliscus. *Bull. Mus. Comp. Zool.*, Harvard Univ. *128* (1): 1–33.

MAYR, E., (1942): Systematics and the origin of species. New York, Columbia Univ. Press.

MAYR, E., (1963): Animal species and evolution. Cambridge, Harvard Univ. Press.

MAYR, E., (1964): Inferences concerning the Tertiary American bird faunas. *Proc. Nat. Acad. Sci. 51*.

MAYR, E., (1967): Artbegriff und Evolution. Hamburg, Berlin.

MAYR, E., (1971): New species of birds described from 1956 to 1965. *J. Orn. 112* (3): 302–316.

MAYR, E., LINSLEY, E. G. and USINGER, R. L., (1953): Methods and principles of systematic zoology. McGraw-Hill, New York.

MAYR, E. and PHELPS, W., (1955): Origin of the bird fauna of Pantepui. In: Acta XI Congressus Internat. Ornith. Basel 1954: 399–400.

MAYR, E. and PHELPS, W., (1967): The origin of the bird fauna of the South Venezuelan Highlands. *Bull. Amer. Mus. Nat. Hist. 136*: 273–327.

MCCOY, C. J., (1963): Eumeces sumichrasti (Reptilia: Scincidae) in Quintana Roo, Mexico. *Herpetologica 15*.

MCCOY, C. J., (1968): A review of the Genus Laemanctus (Reptilia, Iguanidae). *Copeia 1968* (4): 665–678.

MCCOY, C. L., (1969): Snakes of the genus Coniophanes (Colubridae) from the Yucatan Peninsula, Mexico. *Copeia 1969* (4): 847–849.

MCCOY, C. J. and WALKER, CH. F., (1966): A new Salamander of the genus Bolitoglossa from Chiapas. *Occas. Pap. Mus. Zool. Univ. Michigan 649*: 1–11.

MEHELY, L. V., (1904): Investigations on Paraguayan Batrachians. *Ann. Mus. Nat. Hungar. 2*: 207–232.

MELLO-LEITÃO, C. DE, (1936): Distribution et phylogenie des faucheurs sudamericaines. XII Int. Cong. Zool. 2: 1217–1228, Lisbon.

MELLO-LEITÃO, C. DE, (1936): La distribution des Arachnides et son importance pour la zoogeographie Sud-americaine. XII Int. Congr. Zool. 2: 1209–1216, Lisbon.

MELLO-LEITÃO, C. DE, (1939): Les arachnides et la zoogéographie de l'Argentine. *Physis 18*: 601–630.

MELLO-LEITÃO, C. DE, (1942): Los alacranes y la zoogeografia sudamericana. *Rev. Argent. Zoogeogr.* 2: 125–132.

MELLO-LEITÃO, C. DE, (1946): As Zonas de Fauna de América Tropical. *Rev. Bras. Geograf. 1.*

MELLO-LEITÃO, C. DE, (1947): Zoogeografia do Brasil. São Paulo.

MENENDEZ, C. A., (1969): Die fossilen Floren Südamerikas. In: Biogeography and Ecology in South America 2: 519–561.

MERCER, J. H., (1965): Glacier Variations in Southern Patagonia *Geogr. Rev. 55*: 390–413.

MERCER, J. H., (1968): Variations of some Patagonian glaciers since the Late-Glacial. *Americ. Science 266* (2): 91–109.

MERTENS, R., (1924): Ein Beitrag zur Kenntnis der melanistischen Inseleidechsen des Mittelmeeres. *Pallasia* 2: 40–52, Dresden.

MERTENS, R., (1930): Bemerkungen über die Säugetiere der Inseln Lombok, Sumbawa und Flores. Zool. Garten *23.*

MERTENS, R., (1934): Die Inselreptilien, ihre Ausbreitung und Artbildung. *Zoologica 84,* Stuttgart.

MERTENS, R., (1938): Amphibien und Reptilien aus Santo Domingo, gesammelt von Dr. H. BÖKER. *Senckenberg.* 20: 332–342.

MERTENS, R., (1938): Bemerkungen über die brasilianischen Arten der Gattung Liolaemus. *Zool. Anz. 123* (7/9): 220–222.

MERTENS, R., (1939): Herpetologische Ergebnisse einer Reise nach der Insel Hispaniola, Westindien. *Abh. senckenberg. naturf. Ges. 499*: 1–84.

MERTENS, R., (1950): Ein neuer Laubfrosch aus Venezuela. *Senckenbergiana zool. 31* (1/2): 1–2.

MERTENS, R., (1952): Die Amphibien und Reptilien von El Salvador. *Abhdl. Senck. Naturf. Ges. 487.*

MERTENS, R., (1956): Zur Kenntnis der Iguaniden-Gattung Tropidurus in Peru. *Senck. biol. 37* (1/2): 101–134.

MERTENS, R., (1957): Zoologische Beobachtungen im Nebelwalde von Rancho Grande, Venezuela. *Nat. u. Volk 87* (10): 337–344.

MERTENS, R., (1960): Über die Schlangen der Galapagos. *Senck. biol. 41*: 133–141.

MERTENS, R., (1963): Die Wiederentdeckung der Geckonengattung Gonatodes auf den Galapagos. *Senck. biol. 44*: 21–23.

MERTENS, R., (1969): Herpetologische Beobachtungen auf der Insel Tobago. *Salamandra 5* (1/2): 63–70.

MERTENS, R., (1972): Herpetofauna tobagana. *Stutt. Beitr. Naturkd. 252*: 1–22.

MEYER DE SCHAUENSEE, R., (1948–1952): The birds of the Republic of Colombia. *Caldasia 5*: 251–1214.

MEYER DE SCHAUENSEE, R., (1950): Colombian zoological survey; part VII: a collection of birds from Bolivar, Colombia. *Proc. Acad. Nat. Sci. Philad. 102*: 111–139.

MEYER DE SCHAUENSEE, R., (1964): The birds of Colombia and adjacent areas of South and Central America. Narberth, Pennsylvania.

MEYER DE SCHAUENSEE, R., (1966): The species of birds of South America and their distribution. Narberth, Pennsylvania.

MEYER DE SCHAUENSEE, R., (1970): A guide to the birds of South America. Livingston Publ. Comp.

MILA DE LA ROCHE, F., (1932): Introduction al estudio de los ofidios de Venezuela. *Bol. Soc. Ven. Cienc. Nat. Caracas 1* (10): 381–392.

MILLER, A. H., (1947): The tropical avifauna of the upper Magdalena valley, Colombia. *Auk 64*: 351–381.

MILLER, A. H., (1952): Supplementary data on the tropical avifauna of the arid upper Magdalena valley of Colombia. *Auk 69*: 450–457.

MILLER, G. S., (1898): A new Rabbit from Margarita Island, Venezuela. *Proc. Biol. Soc. Washington 12*: 97–98.

MILLER, G. S., (1900): Three new bats from the island of Curacao *Proc. Biol. Soc. Washington 13*: 123–127.

MILLER, R. R., (1966): Geographical distribution of Central American Freshwater Fishes. *Copeia 1966* (4): 773–802.

228

MIROV, N. T., (1967): The Genus Pinus. New York.

MONHEIM, F., (1956): Beiträge zur Klimatologie und Hydrologie des Titicacabeckens. *Heidelberger Geogr. Arb. 1.*

MONOD, T., (1963): The late Tertiary and Pleistocene in the Sahara. In: 'African Ecology and Human Evolution' (F. C. HOWELL and F. BOURLIÈRE, eds), pp. 117–229. Viking Fund Publications in Anthropology, No. 36, New York.

MONROE, B. L., (1968): A Distributional Survey of the Birds of Honduras. Americ. Ornith. Union, Allen Press, Lawrence.

MOREAU, R. E., (1933): Pleistocene climatic changes and their distribution of life in East Africa. *J. Ecol. 21*: 415–435.

MOREAU, R. E., (1963): Vicissitudes of the African biomes in the Late Pleistocene. *Proc. Zool. Soc. London 141*: 395–421.

MOREAU, R. E., (1966): The Bird Faunas of Africa and its Islands. London und New York.

MOREAU, R. E., (1969): Climatic changes and the distribution of forest vertebrates in West Africa. *J. Zool. 158*: 39–61.

MORGAN, H., (1920): Dermaptera and Orthoptera. *Calif. Acad. Sci. Proc. 2 (2)*: 311–346.

MORTELMANS, G. and MONTEYNE, R., (1962): Le quaternaire du Congo occidental et sa chronologie. *Annls. Mus. r. Congo belge, Ser. 8, Vo. 40 (1)*: 97–126.

MORTENSEN, H., (1957): Temperaturgradient und Eiszeitklima am Beispiel der pleistozänen Schneegrenzdepression in den Rand- und Subtropen. *Z. Geomorph. 1*: 44–56.

MOYNIHAN, M., (1962): The organisation and probable evolution of some mixed species flocks of neotropical birds. *Smith. Misc. Coll. 143*: 1–140.

MÜLLER, H., (1971): Ökophysiologische und Ökoethologische Untersuchungen an Cnemidophorus lemniscatus L. (Reptilia: Teiidae) in Kolumbien. *Forma et Functio 4*: 189–224.

MÜLLER, L., (1938): Beiträge zur Kenntnis der Herpetofauna Chiles. X. Über ein Exemplar von Telmatobius montanus Philippi. *Zool. Anz. 121 (11/12)*: 313–317.

MÜLLER, L. and HELLMICH, W., (1932): Beiträge zur Kenntnis der Herpetofauna Chiles. I. Über Borborocoetes kriegi und die Larven einiger chilenischer Anuren. *Zool. Anz. 97*: 7–8, 204–211.

MÜLLER, L. and HELLMICH, W., (1936): Amphibien und Reptilien. I. Teil: Amphibia, Chelonia, Loricata. In: Wissenschaftliche Ergebnisse der Deutschen Gran Chaco-Expedition. Stuttgart.

MÜLLER, L. and HELLMICH, W., (1938): Liolaemus-Arten aus dem westlichen Argentinien. *Zool. Anz. 123*: 130–142. I. Liolaemus darwini (Bell) und Liolaemus goetschi n. sp.

MÜLLER, L. and HELLMICH, W., (1939): Ergebnisse der Argentinienreise von Prof. Dr. W. Goetsch, Breslau. II. Über eine neue Liolaemus altissimus-Rasse vom Volcán Copahue. *Zool. Anz. 125*: 113–119.

MÜLLER, L. and HELLMICH, W., (1939): Liolaemus-Arten aus dem westlichen Argentinien. Ergebnisse der vierten Forschungsreise von Prof. Dr. H. Krieg nach Südamerika. III. Über Liolaemus kriegi, eine neue Lioalemus-Art aus der Gegend des Lago Nahuel Huapi. *Zool. Anz. 127*: 44–47.

MÜLLER, L and HELLMICH, W., (1939): Liolaemus-Arten aus dem westlichen Argentinien. IV. Über Liolaemus-Arten aus den Territorien Rio Negro und Nenquen. *Zool. Anz. 128*: 1–17.

MÜLLER, P., (1966): Studien zur Wirbeltierfauna der Insel von São Sebastião (23°50′S/45°20′W). Saarbrücken.

MÜLLER, P., (1967): Zur Verbreitung von Passer domesticus in Brasilien. *J. Ornith. 108 (4)*: 497–499.

MÜLLER, P., (1968): Die Herpetofauna der Insel von São Sebastião (Brasilien). Verl. Saarbrücker Zeitung, Saarbrücken.

MÜLLER, P., (1968): Einige Bemerkungen zur südamerikanischen Krötenart Bufo granulosus Spix 1824. *datz 21 (8)*: 248–251.

MÜLLER, P., (1968): Zur Verbreitung der Gattung Hydromedusa (Testudines, Chelidae) auf den südostbrasilianischen Inseln. *Salamandra 4*: 16–26.

MÜLLER, P., (1968): Beitrag zur Herpetofauna der Insel Campeche (27°42′S/48°28′W). *Salamandra 4*: 47–55.

MÜLLER, P., (1969): Zur Verbreitung von Hemidactylus mabouia (MOREAU DE JONES) auf den südbrasilianischen Inseln. *Zool. Anz. 182* (3/4): 196–203.

MÜLLER, P., (1969): Zur Verbreitung der Gattung Chironius (Serpentes/Colubridae) auf den südbrasilianischen Inseln. *Senck. biol. 50* (3/4): 133–141.

MÜLLER, P., (1969): Herpetologische Beobachtungen auf der Insel Marajó. *datz 22* (4): 117–121.

MÜLLER, P., (1969): Über den Subspezies-Status südbrasilianischer Bufo ictericus-Populationen. *datz 22* (11): 340–342.

MÜLLER, P., (1969): Über eine neue Subspezies vom Teju, Tupinambis teguixin buziosensis n. ssp. *Salamandra 5* (1/2): 32–35.

MÜLLER, P., (1969): Einige Bemerkungen zur Verbreitung von Vipera aspis (Serpentes/Viperidae) in Spanien. *Salamandra 5* (1/2): 57–62.

MÜLLER, P., (1970): Durch den Menschen bedingte Arealveränderungen brasilianischer Wirbeltiere. *Nat. u. Museum 100* (1): 22–37.

MÜLLER, P., (1970): Vertebratenfaunen brasilianischer Inseln als Indikatoren für glaziale und postglaziale Vegetationsfluktuationen. *Abh. Dtsch. Zool. Ges. Würzburg*, 1969: 97–107.

MÜLLER, P., (1970): Über die Eunectes–Arten von Marajó. *Salamandra 6* (3/4): 140–141.

MÜLLER, P., (1971): Erstnachweis von Caiman crocodylus für die Insel von Florianopolis. *Aquaterra 5*: 59–60.

MÜLLER, P., (1971): Herpetologische Reiseindrücke aus Brasilien. *Salamandra 7* (1): 9–30.

MÜLLER, P., (1971): Beobachtungen an brasilianischen Geochelone carbonaria. *Aquaterra 6*: 69–72, 7.

MÜLLER, P., (1971): Dynamic Zoogeography. *Geoforum 8*: 81–82.

MÜLLER, P., (1971): Erstnachweis von Dermochelys coriacea (L. 1766) für die Insel von São Sebastião (Brasilien). *Salamandra 7*: 86.

MÜLLER, P., (1971): Die Ausbreitungszentren und Evolution in der Neotropis. *Mitt. Biogeogr. Abt. Univ. Saarl. 1*: 1–20.

MÜLLER, P., (1972): Ausbreitungszentren in der Neotropis. *Naturw. Rdsch. 25* (7): 267–270.

MÜLLER, P., (1972): Der neotropische Artenreichtum als biogeographisches Problem. Festbundel Brongersma. *Zool. Mededelingen 47*: 88–110.

MÜLLER, P., (1972): Centres of Dispersal and Evolution in the Neotropical Region. Studies on the Neotropical Fauna.

MÜLLER, P., (1972): Biogéographie et Evolution en Amérique du Sud. Int. Geogr. Congr. Montreal.

MÜLLER, P., (1972): Zweites Symposium über 'Biogeographische und Landschaftsökologische Probleme Südamerikas'. *Mitt. Biogeogr. Abt. Univers. Saarl. 3*.

MÜLLER, P., (1972): Die Bedeutung der Biogeographie für die ökologische Landschaftsforschung. Biogeographica 1.

MÜLLER, P., (1972): Die Bedeutung der Ausbreitungszentren für die Evolution neotropischer Vertebraten. *Zool. Anz.*

MÜLLER, P. and SCHMITHÜSEN, J., (1970): Probleme der Genese südamerikanischer Biota. Festschr. Gentz.

MÜLLER, P. and SCHNEIDER, B., (1969): Bemerkungen zur Systematik und Zoogeographie europäischer Chalcides chalcides (Reptilia/Scincidae) mit besonderer Berücksichtigung der Subspezies vittatus. *Zool. Anz. 182* (5/6): 322–327.

MYERS, CH. W., (1966): The distribution and behavior of a tropical Horned Frog, Cerathyla panamensis STEJNEGER. *Herpetologica 22* (1): 68–71.

MYERS, CH. W., (1969): Snakes of the Genus Coniophanes in Panama. *Amer. Mus. Novitates 2372*: 1–28.

MYERS, CH. W., (1969): The Ecological Geography of Cloud Forest in Panama. *Amer. Mus. Novit. 2396*: 1–52.

MYERS, G. S., (1938): Fresh-water fishes and West Indian zoogeography. *Ann. Rept. Smithsonian Inst.* 339–364.

MYERS, G. S., (1950): The systematic status of Hyla septentrionalis, the large tree frog of the Florida Keys, the Bahamas and Cuba. *Copeia 1950*: 203–214.

MYERS, G. S., (1966): Derivation of the Freshwater Fish fauna of Central America. *Copeia 1966* (4): 766–773.

NADIG, A., (1968): Über die Bedeutung der Massifs de Refuge am südlichen Alpenrand (dargelegt am Beispiel einiger Orthopterenarten). *Mitt. schweiz. ent. Ges. 41*: 341–358.

NEILL W. T., (1965): New and noteworthy amphibians and reptiles from British Honduras. *Bull. Flo. State Mus. 9* (3): 77–130.

NEILL, W. T. and ALLEN, R., (1959): Studies on the amphibians and reptiles of British Honduras. *Publ. Res. Div. Ross Allen's Reptile Inst. Inc. 2*: 1–76.

NEILL, W. T. and ALLEN, R., (1962): Reptiles of the Cambridge Expedition to British Honduras, 1959–60. *Herpetologica 18*: 79–91.

NICEFORO, M. H., (1942): Los ofidios de Colombia. *Rev. Acad. Colomb. Cienc. Bogota 5* (17): 84–101.

NIETHAMMER, G., (1953 and 1956): Zur Vogelwelt Boliviens. *Bonn. Zool. Beitr. 4*: 195–303, 7: 84–150.

NIETHAMMER, J., (1964): Contribution à la connaissance des Mammifères terrestres de l'île Indefatigable (= Santa Cruz), Galapagos. *Mammalia 28* (4): 593–606.

NOBLE, G. K., (1938): A new species of frogs of the genus Telmatobius from Chile. *Amer. Mus. Nov. 973*: 1–3.

NOODT, W., (1965): Natürliches System und Biogeographie der Syncarida (Crustacea/Malacostraca). *Gewäss. Abwäss. 37/38*: 77–186.

NOODT, W., (1967): Biogeographie der Bathynellacea. Proc. Symposium Crustacea, *Mar. biol. Ass. India 31*: 411–417.

NOODT, W., (1969): Die Grundwasserfauna Südamerikas. In: Biogeography and Ecology in South America 2: 659–684.

NOVAES, F. C., (1959): Variação geografica e o problema da especie nas aves do grupo Ramphocelus carbo. *Bol. Mus. Paraense E. Goeldi 22*: 1–63.

NOVAES, F. C., (1964): Uma nova raça geografica de Piprites chloris (Temminck) do Estado do Para (Pipridae, Aves). *Bol. Mus. Goeldi 47*: 45–47.

NOVAES, F. C., (1965): Notas sobre algumas aves de Serra Parima, Territorio de Roraima (Brasil). *Bol. Mus. Goeldi 54*: 1–10.

OLIVARES, A., (1958): Aves de la costa del Pacifico, Municipio de Guapi, Cauca, Colombia. III. *Caldasia 8*: 217–251.

OLIVARES, A., (1962): Aves de la region sur de la Sierra de la Macarena, Meta, Colombia. *Rev. Acad. Colombiana Cienc. Exactas, Fis. y Nat. 11* (44): 305–345.

OLIVER, J. A., (1948): The relationships and zoogeography of the genus Thalerophis Oliver. *Bull. Amer. Mus. Nat. Hist. 92*: 157–280.

OLROG, C. C., (1955): Las Aves Argentinas. Univ. Nac. Tucuman, Inst. Miguel Lillo, Tucuman, 1–343.

OLROG, C. C., (1963): Lista y distribucion de las aves argentinas. Univ. Nac. Tucuman, Inst. Miguel Lillo, Opera *Lilloana 9*: 1–377.

OLROG, C. C., (1969): Birds of South America. In: Biogeography and Ecology in South America, 2: 849–878.

OLSON, E. C. and MCGREW, P. O., (1941): Mammalian fauna from the Pliocene of Honduras. *Bull. Geol. Soc. Amer. 52*: 1219–1244.

OREJAS MIRANDA, B. R., (1958): Dos especies de ofidios nuevos para el Uruguay. *Comunic. Zool. Mus. Hist. Nat. Montevideo 79*: 1–6.

OSCHE, G., (1962): Das Praeadaptationsphänomen und seine Bedeutung für die Evolution. *Zool. Anz. 169* (1/2): 14–49.

OSGOOD, W. H., (1943): The mammals of Chile. *Zool. Ser. Field Mus. 30*.

OSGOOD, W. H. and CONOVER, B., (1922): Game birds from northwestern Venezuela. *Zool. Ser. Field Mus. Nat. Hist. 12* (3): 19–47.

OTERO, J. R. DE, (1941): Notas de una viagem de estudos aos campos do sul de Mato Grosso. Min. Agr., Rio de Janeiro.

OTREMBA, E., (1954): Südlich des Orinoco. *Die Erde 6*: 147–166.

PAFFEN, K. H., (1957): Caatinga, Campos und Urwald in Ostbrasilien. Dt. Geogr., Hamburg.

PALMÉN, E., (1944): Die Anemohydrochore Ausbreitung der Insekten als zoogeographischer Faktor. *Ann. Zool. Sc. Zool. Bot. Fenn. Vanamo 10* (1): 1–262.

PARKER, H. W., (1933): A list of the frogs and toads from Trinidad. *Trop. Agr.* (Trinidad) *10* (1): 8–12.

PARKER, H. W., (1935): The Frogs, Lizards and Snakes of British Guiana. *Proc. Zool. Soc. London.*

PARKER, H. W., (1936): A collection of reptiles and amphibians from the Upper Orinoco. *Bull. Mus. Roy. Hist. Nat. Belgique 12* (26): 1–4.

PARKER, H. W., (1938): The Vertical distribution of some reptiles and amphibians in southern Ecuador. *Ann. Mag. Nat. Hist. 2* (11): 438–450.

PARODI, L. R., (1930): Ensayo fitogeografico sobre el partido de Pergamino. Rev. Fac. *Agron y Veterin Entr.* I, 7: 65–261. Buenos Aires.

PARODI, L. R., (1940): Los Talares en la provincia de Buenos Aires. *Darwiniana 4*: 33–56, Buenos Aires.

PARODI, L. R., (1940): Los bosques naturales de la provincia de Buenos Aires. *An Acad. bac. Ci. econ. fis. y nat. 7*: 79–90, Buenos Aires.

PARODI, L. R., (1942): Porque no existen bosques naturales en la Llanura bonariense si los arboles crecen en ella cuando se los cultiva? Agronomia, *Rev. Centro Estud. de Agron.* Buenos Aires *30*: 387–390.

PARODI, L. R., (1947): La estepa pampeana. In: Geografia de la Republica Argentina *8*: 143–207. Buenos Aires.

PATTERSON, B. and PASCUAL, R., (1968): Evolution of mammals on southern continents. V. The fossil mammal fauna of South America. *Quart. Rev. Biol. 43*: 409–451.

PAYNTER, R. A., (1955): The ornithogeography of the Yucatan Peninsula. Peabody Mus. *Nat. Hist. Bull. 9*: 1–347.

PEARSON, O. P., (1948): Life history of mountain viscachas in Peru. *J. Mammal. 29*: (4) 345–374.

PEARSON, O. P., (1951): Mammals of the highlands of southern Peru. *Bull. Mus. Comp. Zool. 106* (3): 117–174.

PEARSON, O. P., (1954): Habits of the Lizard Liolaemus multiformis multiformis at High Altitudes in Southern Peru. *Copeia 1954* (2): 111–116.

PEARSON, O. P., (1957): Additions to the mammalian fauna of Peru and notes on some other Peruvian mammals. *Breviora 73*: 1–7.

PEARSON, O. P., (1960): Biology of the subterranean rodents, Ctenomys, in Peru. Mem. *Museo Hist. Natur. Javier Prado 9*: 1–55.

PEBERDY, P. S., (1941): Ornithological collection from Mt. Roraima. *Repts. British Guiana Mus. and Georgetown Pub. Library,* 30–37.

PELLEGRIN, J., (1909): Description de cinq Lézards nouveaux des Hauts-Plateaux du Pérou et de la Bolivie, appartenant au genre Liolaemus. *Bull. Mus. Nat. Hist. Nat. 15*: 324–329.

PENARD, F. P. and PENARD, A. P., (1908–10): De Vogels van Guyana (Suriname, Gayenne en Demarara). Paramaribo.

PERACCA, G. M., (1895): Rettili ed Anfibi, in: Viaggio del dott. Alfredo Borelli nella Republica Argentina y nel Paraguay. X. *Bol. Mus. Zool. Anat. Comp. Torino 10*: 1–32.

PERACCA, G. M., (1914): Reptiles et Batrachiens de Colombie. *Mem. Soc. Sci. Neuchatel 5*: 96–111.

PEREYRA, J. A., (1937): Aves de la Pampa. *Mem. Jardin Zool. 7.*

PEREYRA, J. A., (1951): Avifauna Argentina. *Hornero 9.*

PETERS, J., (1931–1962): Check-list of birds of the world. *Mus. Comp. Zool. 1–7, 9, 15.*

PETERS, J. A., (1953): Snakes and lizards from Quintana Roo, Mexico. *Lloydia 16*: 227–232.

PETERS, J. A., (1960): The snakes of the subfamily Dipsadinae. Misc. *Publ. Mus. Zool. Univ. Michigan 114*: 1–224.

PETERS, J. A., (1960): The snakes of Ecuador. A check list and key. *Bull. Mus. Comp. Zool. 122*: 491–541.

PETERS, J. A. and OREJAS-MIRANDA, B., (1970): Catalogue of the Neotropical Squamata: Part I. Snakes; Part II. Lizards and Amphisbaenians. Smiths. Inst. 297.

PETERS, W., (1877): Sammlung des Hr. Dr. Carl Sachs in Venezuela. Monatsb. Akad. Wissensch. Berlin, 457–460.

232

PHELPS, W., (1938): La procedencia geografica de las aves coleccionadas en el Cerro Roraima. *Bol. Soc. Ven. Cienc. Nat. 5* (36): 2657–7038.

PHELPS, W., (1943): Las aves de Perija. *Bol. Soc. Ven. Cienc. Nat. 8* (56): 265–338.

PHELPS, W. and PHELPS, W., (1958–1963): Lista de las aves de Venezuela. *Bol. Soc. Venez. Cienc. Nat. 19* (90): 1–317, *12* (75): 1–427, *24* (104): 1–179.

PHELPS, W. and PHELPS, W., (1962): Cuarentinueva aves nuevas para la avifauna Brasilera del Cerro Ueitepui (Cerro del Sol). *Bol. Soc. Venez. Cienc. Nat. 23* (101): 32–39.

PHILIPPI, R. A., (1902): Suplemento a los batraquios chilenos descritos en la Historia Fisica y Politica de Chile de don Claudio Gay. Santiago.

PHILIPPI, R. A., (1964): Catalogo de las aves Chilenas. *Inv. Zool. Chil. 11*: 1–179.

PIFANO, F., (1935): Contribucion al Estudio de las Serpientes ponzonosas del Estado Yaracuy. Caracas 1–16.

PIFANO, F., (1938): Corales ponzonosas de los Valles del Yaracuy. Pesquisas experimentales con la ponzona del Micrurus lemniscatus (L. 1758). *Publ. Asoc. Med. Yaracuy 1*: 10–15.

PIFANO, F., (1954): Emponzonamientos producidos por serpientes ponzonosas venezolanas. Caracas.

PIFANO, F. and ROMER, M., (1949): Ofidos ponzonosas de Venezuela. I. Nueva comprobacion en Venezuela de Bothrops medusa (STERNFELD 1920) Amaral 1929 y redescription de la especie. *Arch. Venez. Patol. Trop. Paras. Med. 1* (2): 277–300.

PIFANO, F. and ROMER, M., (1949): Ofidios ponzonosos de Venezuela. II. Sobre las serpientes ponzonosas venezolanas del grupo de Bothrops lansbergii. *Arch. Ven. Pat. Trop. Paras. Med. 2*: 301–326.

PIFANO, F., ROMER, M, and SANDNER, M., (1950): Serpientes ponzonosas de Venezuela. III. Bothrops schlegelii (Berthold, 1846) Jan 1875: su existencia en Venezuela. *Arch. Ven. Patol. Trop. Paras. Med. 2* (2): 257–264.

PIMIENTA, J., (1958): A faixa costeira meridional de Santa Catarina, Brasil. *Bol. Div. Geol. Mineral. Brasil 176*: 1–104.

PINTO, O. M. DE O., (1932): Resultados ornithologicos de uma excurção pelo oeste de São Paulo e sul de Mato Grosso. *Rev. Mus. Paul. 19.*

PINTO, O. M. DE O., (1938–1944): Catalogo das aves do Brasil. *1*, *Rev. Mus. Paul. 22*: 2. Dep. Zool. Sec. Agric. Ind. Comerc., São Paulo.

PINTO, O. M. DE O., (1940): Nova contribuição a ornithologia de Mato Grosso. *Arq. Zool. 2* (1), São Paulo.

PINTO, O. M. DE O., (1947): Contribuição a ornithologia de Mato Grosso. *Arq. Zool. 5* (6), São Paulo.

PINTO, O. M. DE O., (1949): Conçeito atual e nomenclatura revista das aves alistadas no 'Catalogo' de E. Snethlage. *Bol. Mus. Goeldi 10*: 1–80.

PINTO, O. M. DE O., (1964): Ornitologia Brasiliense. Dep. Zool. Secr. Agric. Estado de São Paulo, São Paulo.

POLANSKI, J., (1965): The maximum glaciation in the Argentine Cordillera. International Studies of the Quaternary. *Geol. Soc. Amer. 84*: 453–472.

PORTER, C., (1898): Contribucion a la fauna de la Provincia de Valparaiso. *Rev. Chil. Hist. Nat. 2*: 31–33.

PRADO, A., (1945): Notas ofidiologicas. 19. Atractus da Colombia, com a redescrição de tres novas especies. *Mem. Inst. But. 18*: 109–114.

PUTZER, H., (1958): Quatäre Krusten-Bildungen im tropischen Südamerika. *Geol. Jb. 76*: 37–52.

PUTZER, H., (1968): Überblick über die Geologische Entwicklung Südamerikas. In: Biogeography and Ecology in South America. *1*: 1–24.

QUELCH, J. J., (1899): The Poisonous Snakes of British Guiana. *Ann. Mag. Nat. Hist. 7* (3): 402–409.

QUIJADA, B., (1911): Principales Rasgos de la Geografia Animal de Chile (1). *Bol. Mus. Nac. Chile, 3* (1): 146–151.

QUIJADA, B., (1914): Catalogo de los Batracios chilenos y estranjeros conservados en el Museo Nacional. *Bol Mus. Nac. Chile 7* (1): 319–326.

233

RAMBO, B., (1946): A fisionomia do Rio Grande – Viagens de estudos. *Bol. Geogr. 40*: 410–424, *41*: 555–569.

RAMBO, B., (1948): A flora austral antarctica e andina no Rio Grande do Sul. *Bol. Geogr. 67*: 750–754, Rio de Janeiro.

RAMBO, B., (1956): A fisionomia do Rio Grande do Sul. Ensaie de monografia natural. Porto Alegre.

RAPOPORT, E. H., (1968): Algunos problemas biogeograficos del Nuevo Mundo con especial referencia a la region neotropical. In: Biologie de l'Amérique australe, 55–110, Paris.

RAWITSCHER, F. K., (1948): The water economy of the vegetation of the 'Campos' in southern Brazil. *J. Ecology 36*.

RAWITSCHER, F. K., (1950): Climax and Pseudoclimax Vegetation in the Tropics (South America). Proc. seventh intern. Bot. Congr. Stockholm.

RAWITSCHER, F. K., HUECK, K., MORELLO and PAFFEN, K. H., (1952): Algunas observações sobre a ecologia da vegetação das Caatingas. *An. Acad. Bras. Cienc. 24*.

REIG, O., (1968): Peuplement en vertébrès tétrapodes de l'Amérique du Sud. In: Biologie de l'Amérique australe. Paris.

REINIG, W. F., (1937): Die Holarktis. Jena.

REINIG, W. F., (1950): Chorologische Voraussetzungen für die Analyse von Formenkreisen. Syllegomena biologica. Festschr. O. Kleinschmitt, 346–378, Leipzig.

REINKE, R., (1962): Das Klima Amazoniens. Dissertation, Tübingen.

REMANE, A., (1961): Gedanken zum Problem: Homologie und Analogie, Praeadaptation und Parallelität. *Zool. Anz. 166*: 447–465.

RENDAHL, H., (1918–1920): Notes on a collection of birds from Panama, Costa Rica and Nicaragua. *Ark. Zool. 12* (8): 1–36.

REYNE, A., (1921): Over het voorkomen van den Rotshaan (Rupicola crocea L.) in Suriname. *Ardea 10*: 25–26.

RICHARDS, P. W., (1964): The Tropical Rain Forest. Cambridge Univ. Press.

RILEY, H. P., (1952): Ecological barriers. *Amer. Nat. 86*.

RINGUELET, R. A., (1955): Vinculaciones faunisticas de la zona boscosa del Nahuel Huapi y el dominio zoogeografico Australcordillerano. *Notas Mus. La Plata 18*: 81–121.

RINGUELET, R. A., (1961): Rasgos fundamentales de la zoogeografia de la Argentina. *Physis 22*: 151–170.

RIVAS, L. R., (1958): The origin, evolution, dispersal, and geographical distribution of the Cuban poeciliid fishes of the tribe Girardinini. *Proc. Amer. Philos. Soc. 102* (3): 281–320.

RIVERO, J., (1961): Salientia of Venezuela. *Bull. Mus. Comp. Zool. 126* (1): 1–207.

RIVERO, J., (1964): The distribution of Venezuelan frogs. VI. The Llanos and the delta region. *Caribb. J. Sci. 4*: 491–495.

ROBINSON, H., (1965): Venezuelan bryophytes collected by Julian A. Steyermark. *Acta. Bot. Venezuelica 1*: 73–83.

ROESLER, U., (1965): Chorologische Untersuchungen über den Homoeosoma-Ephestia-Komplex (Lepidoptera: Phycitinae) im paläarktischen Raum. *Bonn. zool. Beitr. 16*.

ROHDENBURG, H., (1970): Morphodynamische Aktivitäts- und Stabilitätszeiten statt Pluvial- und Interpluvialzeiten. *Eiszeitalter und Gegenwart 21*: 81–96.

ROHL, E., (1949): Fauna descriptiva de Venezuela, Vertebrados. *Bol. Acad. Cienc. Fis. Nat. Mat. 12* (36/37).

ROMER, A. S., (1966): Vertebrate Paleontology. Chicago und London.

ROOY, N. DE, (1922): Reptiles and amphibians of Curaçao. *Bijdr. Dierk. 22*: 249–253.

ROSEVAERE, G. M., (1948): The grassland of Latin America. Imp. Bur. Pastures and Field Crops *356*, Aberytwyth.

ROUX, J., (1910): Eine neue Cystignathidenart aus Chile. *Zool. Anz. 36*.

ROUX, J., (1927): Contribution à l'erpétologie du Venezuela. *Verh. Naturf. Ges. Basel 38*: 252–261.

ROZE, J. A., (1952): Contribucion al conocimiento del estudio de las familias Typhlopidae y Leptotyphlopidae en Venezuela. *Mem. S. C. N. La Salle 12* (32): 143–158.

ROZE, J. A., (1952): Coleccion de reptiles del prof. Scorza de Venezuela. *Acta Biol. Venez. 1* (5): 93–114.

234

ROZE, J. A., (1953): Ofidios de Camuri Chico, Macuto, D. F. Venezuela colectados por el Rvdo. Padre Cornelius Vogl. *Bol. Soc. Ven. Cienc. Nat. Caracas 14* (79): 200–211.

ROZE, J. A., (1953): The Rassenkreis Coluber (Masticophis) mentovarius (D., B. u. D., 1854). *Herpetologica 9* (3): 113–120.

ROZE, J. A., (1954): Notas preliminar sobre los ofidios de la expedicion Franco-Venezolana al Alto Orinoco. *Arch. Venez. Patol. Trop. Paras. Med. 2* (2): 227–237.

ROZE, J. A., (1955): Revision de las corales del genero Micrurus (Serpentes: Elapidae) de Venezuela. *Acta Biol. Venez. 1* (17): 453–500.

ROZE, J. A., (1955): Ofidios coleccionados por la Expedicion Franco-Venezolana al Alto Orinoco 1951 a 1952. *Bol. Mus. Cienc. Nat. Caracas 1* (3/4): 179–195.

ROZE, J. A., (1958): Resultados zoologicos de la Expedicion de la Universidad Central de Venezuela a la region del Auyantepui en la Guayana venezolana, abril de 1956. 5. Los reptiles del Auyantepui, Venezuela, basandose en las colecciones de las expediciones de Phelps-Tate, del American Museum of Natural History 1937–1938 y de la Universidad Central de Venezuela 1956. *Acta Biol. Venez. 2* (22): 243–270.

ROZE, J. A., (1958): A new species of the genus Urotheca (Serpentes: Colubridae) from Venezuela. *Breviora 88*: 1–5.

ROZE, J. A., (1958): Los reptiles del Chimanta Tepui (Estado Bolivar) Venezuela, colectados por la expedition botanica del Chicago Natural History Museum. *Acta Biol. Venez. 2* (25): 299–314.

ROZE, J. A., (1959): Una nueva especie del genero Drymarchon (Serpentes: Colubridae) de la isla Margarita, *Venezuela. Noved. Cient. La Salle, Caracas, ser. zool. 25*: 1–4.

ROZE, J. A., (1959): Taxonomic notes on a collection of Venezuelan reptiles in the American Museum of Natural History. *Amer. Mus. Novitates 1934*: 1–4.

ROZE, J. A., (1959): El genero Erythrolamprus Wagler (Serpentes: Colubridae) en Venezuela. *Acta Biol. Venez. 2* (35): 523–534.

ROZE, J. A., (1961): El genero Atractus (Serpentes: Colubridae) en Venezuela. *Acta Biol. Venez. 3* (7): 103–119.

ROZE, J. A., (1966): La Taxonomia y Zoogeografia de los ofidos en Venezuela. Caracas.

ROZE, J. A. and TREBBAUM, C. P. M., (1958): Un nuevo genero de corales venenosas (Leptomicrurus) para Venezuela. *Acta Cient. Venez. 9*: 128–130.

RÜHM, W., (1969): Die Nematoden der an Araucaria araucana (MOL.) KOCH und Araucaria angustifolia KUNTZE gebundenen Scolytoidea (Col.) und ihre verwandtschaftliche Stellung zur Nematodenfauna der paläarktischen Borkenkäfer. *Beitr. Neotr. Fauna 6* (2): 137–144.

RUTHVEN, G., (1914): Description of a new species of Brasiliscus from the region of the Sierra Nevada de Santa Marta, Colombia. *Proc. Biol. Soc. Washington 27*: 9–12.

RUTHVEN, G., (1923): The reptiles of the Dutch Leeward Islands. *Occ. Pap. Mus. Zool. Michigan 143*.

RUTHVEN, G., (1924): Description of an Ameiva from Testigos Island, Venezuela. *Occ. Pap. Mus. Zool. Michigan 149*.

RUTTEN, L. M. R., (1932): De geologische geschiedenis der drie Nederlandsche Benedenwindse Eilanden. *De West-Indische Gido 13*: 401–441.

RUTTEN, M. G., (1932): Verordening tot Bescherming van Vogels op Curaçao. *Ardea 21*: 73–74.

RYAN, R. M., (1963): The biotic provinces of Central America. *Acta Zool. Mex. 6* (2/3): 1–55.

SACCONE, R., (1962): Contribucion al conocimiento de los charadriiformes del Uruguay. *Comunic. Socied. Zool. Uruguay.*

SAHLSTEIN, T. G., (1932): Petrologie der postglazialen vulkanischen Aschen Feuerlands. *Acta Geogr. 5* (1).

SALMI, M., (1941): Die postglazialen Eruptionsschichten Patagoniens und Feuerlands. *Ann. Acad. Sc. Fenn.*, Helsinki.

SAPPER, K., (1937): Mittelamerika, in Handbuch der Regionalen Geologie 29 (8): 1–160.

SAUTER, W., (1968): Zur Zoogeographie der Schweiz am Beispiel der Lepidopteren. *Mitt. schweiz. ent. Ges. 41*: 330–336.

SAVAGE, D. E., (1951): Report on fossil vertebrates from the upper Magdalena Valley, Colombia. *Science 114*: 186–187.

SAVAGE, J. M., (1966): The origins and history of the Central American Herpetofauna. *Copeia 1966* (4): 719–766.

SAVAGE, J. M., (1968): Evolution. BLV. Bayer. Landwirtschaftsverl., München, Basel, Wien.

SAVAGE, J. M., (1968): The Dendrobatid frogs of Central America. *Copeia 1968* (4): 745–776.

SAVAGE, J. M., (1968): The distribution and synonymy of the neotropical frog, Eleutherodactylus moro *Copeia 1968* (4): 878–879.

SAVAGE, J. M., (1969): Clarification of the status of the toad Bufo veraguensis O. Schmidt. *Copeia 1969* (1): 178–179.

SAVAGE, J. M. and HEYER, W. R., (1967): Variation and distribution in the three-frog genus Phyllomedusa in Costa Rica, Central America. *Beitr. Neotr. Fauna 5* (2): 111–131.

SCHAEFFER, B., (1949): Anurans from the early tertiary of Patagonia. *Bull. of the Amer. Nat. Hist. Mus. 93* (2): 47–68.

SCHÄFER, E., (1969): Über Lebensweise und Ökologie der im Nationalpark von Rancho Grande (Nord-Venezuela) nachgewiesenen Ameisenvogelarten (Formicariidae). *Bonner Zool. Beitr. 1/3*: 99–109.

SCHALLER, F., (1958): Bilder aus den Lebensräumen Perus. *Verhdl. Dtsch. Zool. Ges. 21*: 536–537, Graz.

SCHARFF, R. F., (1922): On the origin of the West Indian fauna. Bydr. Dierk., 65–72, Amsterdam.

SCHMIDT, K. P., (1932): Reptiles and amphibians of the Mandel Venezuelan Expedition. *Zool. Ser. Field Mus. Nat. Hist. 18* (7): 157–163.

SCHMIDT, K. P., (1936): Notes on Central American and Mexican coral snakes. *Zool. Ser. Field Mus. Nat. Hist. 20*: 205–216.

SCHMIDT, K. P., (1939): A new Coral Snake from British Guiana. *Zool. Ser. Field Mus. Nat. Hist. 24* (6): 45–47.

SCHMIDT, K. P., (1952): A new Leptodactylid frog from Chile. *Fieldiana Zool. 34* (2): 11–15.

SCHMIDT, K. P., (1954): Notes on frogs of the genus Telmatobius. *Fieldiana Zool. 34* (26): 277–287.

SCHMIDT, K. P., (1954): Reports of the Lund University Chile Expedition 1948/49, 13, Amphibia Salientia. *Lunds Univ. Arskrift.*, N. F., *2*.

SCHMIDT, K. P., (1955): Coral snakes of the genus Micrurus in Colombia. *Fieldiana Zool. 34*(34): 337–358.

SCHMIDT, K. P., (1957): The venomous coral snakes of Trinidad. *Fieldiana Zool. 39* (8): 55–63.

SCHMIDT, K. P. and ANDREWS, E. W., (1936): Notes on snakes from Yucatan. *Zool. Ser. Field Mus. Nat. Hist. 20*: 167–187.

SCHMIDT, R. D., (1952): Die Niederschlagsverteilung im andinen Kolumbien. *Bonner Geogr. Abhdl. 9*: 99–119.

SCHMIEDER, O., (1927): The pampa, a natural or culturally induced grassland? – *Univ. California Publ. in Geogr. 2*: 255–270.

SCHMITHÜSEN, J., (1953): Die Grenzen der chilenischen Vegetationsgebiete. Dt. Geogr. Essen.

SCHMITHÜSEN, J., (1954): Waldgesellschaften im nördlichen Mittelchile. *Vegetatio 5/6*.

SCHMITHÜSEN, J., (1954): Immergrüne Hartlaubgehölze des subtropischen Winterregengebietes in Chile. Rhododendron Jb., Bremen.

SCHMITHÜSEN, J., (1956): Die räumliche Ordnung der chilenischen Vegetation. *Bonner Geogr. Abh. 17*: 1–86.

SCHMITHÜSEN, J., (1957): Probleme der Vegetationsgeographie. Dtsch. Geogr.

SCHMITHÜSEN, J., (1960): Die Nadelhölzer in den Waldgesellschaften der südlichen Anden. *Vegetatio 9*: 313–327, Den Haag.

SCHMITHÜSEN, J., (1966): Problems of Vegetation history in Chile and New Zealand. *Vegetatio 13*.

SCHMITHUSEN, J., (1968): Allgemeine Vegetationsgeographie. Berlin.

SCHMUCKER, TH., (1942): Silvae Orbis. Berlin-Wannsee.

SCHNEIDER, C. O., (1930): Observaciones sobre batracios chilenos. *Rev. Chil. Hist. Nat. 34*: 220–223.

SCHOMBURG, R. K., (1840): Journey from Fort San Joaquim, on the Rio Branco, to Raraima, and thence by the rivers Parima and Meriroari to Esmeralda, on the Orinoco, in 1838–1839. *J. Roy. Geogr. Soc. London 10*: 191–267.

236

SCHOMBURG, R. K., (1848): Versuch einer Fauna und Flora von British-Guiana. In Schomburg Richard, Reisen in Britisch-Guiana in den Jahren 1840–1844. Leipzig *3*: 533–1260.

SCHÖNFELD, G., (1947): Hölzer aus dem Tertiär von Kolumbien. *Abhdl. Senckenb. naturf. Ges. 475*: 1–53.

SCHUCHERT, C., (1935): Historical geology of the Antillean-Caribbean region. New York.

SCHWABE, G. H., (1956): Die ökologischen Jahreszeiten im Klima von Mininco (Chile). *Bonner Geogr. Abh. 17.*

SCHWARTZ, A., (1960): The large toads of Cuba. *Proc. Biol. Soc. Washington 73*: 45–56.

SCHWARTZ, A., (1964): Anolis equestris in Oriente, Cuba. *Bull. Mus. Comp. Zool., Harvard Univ. 131* (12): 406–425.

SCHWARTZ, A., (1965): Variation and Natural History of Eleutherodactylus ruthae on Hispaniola. *Bull. Mus. Comp. Zool., Harvard Univ. 132* (6): 479–508.

SCHWARTZ, A., (1966): The relationships of four small Hispaniolan Eleutherodactylus (Leptodactylidae). *Bull. Mus. Comp. Zool., Harvard Umiv., 133* (8): 369–399.

SCHWARTZ, A., (1966): The Ameiva (Reptilia, Teiidae) of Hispaniola. I. Ameiva lineolata Duméril and Bibron. *Carib. Jour. Sci. 5* (1/2): 45–57.

SCHWARTZ, A., (1966): The Leiocephalus (Lacertilia, Iguanidae) of Hispaniola. I. Leiocephalus melanochlorus Cope. *Jour. Ohio Herp. Soc. 5* (2): 39–48.

SCHWARTZ, A., (1967): The Ameiva (Lacertilia, Teiidae) of Hispaniola. III. Ameiva taeniura Cope. *Bull. Mus. Comp. Zool. 135* (6): 345–375.

SCHWARTZ, A. and MCCOY, CL. J., (1970): A systematic review of Ameiva auberi COCTEAU (Reptilia, Teiidae) in Cuba and the Bahamas. *Am. Carnegie Mus. 41*: 45–168.

SCHWEIGGER, E., (1959): Die Westküste Südamerikas im Bereich des Perustroms. Keysersche Verlagsbuchhdl., Heidelberg, München.

SCHWEINFURTH, CH., (1967): Orchidaceae of the Guayana Highland. *Mem. New York Bot. Garden 14*: 69–214.

SCLATER, P. L., (1858): On the general geographical distribution of the members of the class Aves. *J. Proc. Linn. Soc. London (Zool.) 2*: 130–145.

SCLATER, W. L. and SCLATER, P. L., (1899): The geography of mammals. London.

SEARS, P. B., FOREMAN, F. and CLISBY, K. H., (1955): Palynology in southern North America. *Bull. Geol. Soc. Amer. 66*: 471–530.

SELANDER, R. K. and JOHNSTON, R. F., (1967): Evolution in the house sparrow. I. Intrapopulation variation in North America. Condor *69*: 217–258.

SEXTON, O. J. and HEATWHOLE, H., (1965): Life history notes on some Panamanian snakes. *Caribb. J. Sci. 5*: 39–45.

SHREVE, B., (1947): On Venezuelan reptiles and amphibians collected by Dr. H. G. KUGLER. *Bull. Mus. Comp. Zool. 99* (5): 517–537.

SHREVE, B. and GANS, C., (1958): Thamnophis bovallii DUNN rediscovered (Reptilia, Serpentes). *Breviora 83*: 1–8.

SIBLEY, CH. G., (1958): Hybridization in some Colombian tanagers, avian genus Ramphocelus. *Proc. Amer. Phil. Soc. 102*: 448–453.

SICK, H., (1965): A fauna do Cerrado. *Arch. Zool., São Paulo 12*: 71–93.

SICK, H., (1966): As Aves do Cerrado como Fauna arboricola. *Acad. Brasil. Ciencies 38* (2): 355–363.

SICK, H., (1967): Coryophaspiza melanotis marajoara subsp. nov. *J. Ornith. 108* (2): 218–220.

SICK, H., (1969): Über einige Töpfervögel (Furnariidae) aus Rio Grande do Sul, Brasilien, mit Beschreibung eines neuen Cinclodes. *Beitr. Neotr. Fauna 6* (2): 63–79.

SIMPSON, G. G., (1940): Review of the mammal-bearing Tertiary of South America. *Proc. Amer. Philos. Soc. 83* (5): 649–709.

SIMPSON, G. G., (1950): History of the fauna of Latin America. *Amer. Scientist 38*: 361–389.

SIMPSON, G. G., (1956): Zoogeography of West Indian Land Mammals. *Amer. Mus. Novitates 1759*: 1–28.

SIMPSON, G. G., (1965): The geography of Evolution. Philadelphia, New York.

SIMPSON, G. G., (1966): Mammalian evolution on the southern continents. *Neues Jb. Geol. Paläontol. 125*: 1–18.

237

SIMPSON, G. G., (1969): South American Mammals. In: Biogeography and Ecology in South America 2: 879–909.
SIOLI, H., (1967): The Cururu Region in Brasilian Amazonia. A Transition Zone between Hylaea and Cerrado. *Journ. Indian Bot. Soc.* XLVI (4): 452–462.
SIOLI, H., (1968): Zur Ökologie des Amazonasgebietes. In: Biogeography and Ecology in South America *I*: 137–170.
SKOTTSBERG, C., (1942): The Falkland Islands. *Chronica Botanica 7*: 23–26.
SKUTCH, A. F., (1944): The life history of the Prong-billed Barbet. *Auk* 61: 61–88.
SKUTCH, A. F., (1954): Life histories of Central American birds I. *Pacific Coast Avif.* 31: 1–448.
SKUTCH, A. F., (1958): Roosting and nesting of araçari toucans. *Condor* 60: 201–219.
SKUTCH, A. F., (1960): Life histories of Central American birds. II. *Pacific Coast Avif.* 34:1–593.
SKUTCH, A. F., (1964): Life history of the Scaly-breasted Hummingbird. *Condor* 66: 186–198.
SKUTCH, A. F., (1964): Life histories of Central American pigeons. *Wilson Bull.* 76: 211–247.
SKUTCH, A. F., (1967): Life histories of Central American highland birds. *Nutt. Ornith. Club 7*.
SLUD, P., (1957): The song and dance of the Longtailed Manakin, Chiroxiphia linearis. *Auk, 74*: 333–339.
SLUD, P., (1964): The Birds of Costa Rica. *Bull. Amer. Mus. Nat. Hist.*, 128.
SMITH, E. E., (1954): The forests of Cuba. *Maria Moors Cabot Found 2*: 1–98.
SMITH, H. M., (1938): Notes on reptiles and amphibians from Yucatan and Campeche Mexico. *Occ. Papers Mus. Zool. Univ. Michigan 388*: 1–22.
SMITH, H. M., (1947): Notes on Mexican amphibians and reptiles. *J. Washington Acad. Sci. 37*: 408–412.
SMITH, H. M. and TAYLOR, E. H., (1945): An annotated checklist and key to the snakes of Mexico. *U.S. Nat. Mus. 187*.
SMITH, H. M. and TAYLOR, E. H., (1948): An annotated checklist and key to the Amphibia of Mexico. *U.S. Nat. Mus. Bull. 194*.
SMITH, H. M. and TAYLOR, (1950): An annotated checklist and key to the reptiles of Mexico exclusive of the snakes. *U.S. Nat. Mus. Bull. 199*.
SMITH, L. B., (1967): Bromeliaceae of the Guayana Highland. *Mem. New York Bot. Garden 14*: 15–68.
SNETHLAGE, E., (1906/1907): Über unteramazonische Vögel. *J. Ornith. 54*: 407–411, 55: 283–299.
SNETHLAGE, E., (1910): Sobre a distribuição da avifauna campestre na Amazonia. Bol Mus. Goeldi *6*: 226–236.
SNETHLAGE, E., (1913): Über die Verbreitung der Vogelarten in Unteramazonien. *J. Ornith.61*: 469–539.
SNYDER, D. E., (1966): The birds of Guyana. Salem. Mass., 1–308.
SOLANO, H., (1969): Beiträge zur Kenntnis der Amphibienfauna Venezuelas. *Veröff. Zool. Staatssamml. München 13*: 1–26.
SPASSKY, B., RICHMOND, R. C., PEREZ-SALAS, S., PAVLOVSKY, O., MOURAO, C. A., HUNTER, A. S., HOENIGSBERG, H., DOBZHANSKY, TH. und AYALA, F. J., (1971): Geography of the Sibling Species related to Drosophila willistoni, and of the Semispecies of the Drosophila paulistorum complex. *Evolution 25* (1): 129–14..
STANLEY, D. J., (1970): The ten-fathom terrace on Bermuda: its significance as a datum for measuring crustal mobility and eustatic sea-level changes in the Atlantic. *Z. f. Geomorphol. 14* (2): 186–201.
STEINBACHER, J., (1962): Beiträge zur Kenntnis der Vögel Paraguays. *Abhdl. Senckenb. Natur. Ges. 502*: 1–106.
STEINBACHER, J., (1968): Weitere Beiträge über Vögel von Paraguay. *Senck. biol. 49* (5): 317–365.
STEJNEGER, L., (1901): An annotated list of batrachians and reptiles collected in the vicinity of La Gueira, Venezuela, with descriptions of two new species of snakes. *Proc. U.S. Nat. Mus. 24*: 179–192.
STEJNEGER, L., (1941): Notes on Mexican turtles of the genus Kinosternon. *Proc. U.S. Nat. Mus. 90*: 457–459.

STEVEN, D. M., (1953): Recent evolution in the genus Clethrionomys. Sympos. Soc. Exptl. Biol. 7: 310–319.

STEVENSON, F. J. and CHENG, C. N., (1969): Amino acid levels in the Argentine Basin sediments. Correlation with Quaternary climatic changes. *J. Sed. Petrol. 1969*: 345–349.

STEWART, A., (1911): A botanical survey of the Galapagos Islands. *Proc. Calif. Acad. Sci. 1*: 7–288.

STEYERMARK, J. A., (1962): Botanical novelties in the region of Sierra de Lema, Estado Bolivar. *Soc. Venez. Cien. Nat. 22*: 291–297.

STEYERMARK, J. A., (1962): Botanical novelties from upper Rio Paragua, Estado Bolivar, Venezuela. *Soc. Venez. Cien. Nat. 23*: 89–91.

STEYERMARK, J. A., (1966): Contribuciones a la flora de Venezuela. *Acta Bot. Venezuelica 1* (3/4): 9–256.

STIMPSON, F. A., (1969): Liste der rezenten Amphibien und Reptilien. Boidae. *Das Tierreich 89*: 1–49.

STIRTON, R. A., (1950): Late Cenozoic avenues of dispersal for terrestrial animals between North America and South America. *Bull. Geol. Soc. America 61*: 1541–1542.

STIRTON, R. A., (1954): Late Miocene mammals from Oaxaca, Mexico. *Amer. J. Sci. 252*: 634–638.

STRESEMANN, E., (1927–1934): Aves. In: Hdb. Zool. 7.

STRESEMANN, E., (1939): Die Vögel von Celebes (I–II). *J. Ornith. 87*: 299–425.

STUART, L. C., (1935): A contribution to a knowledge of the herpetology of a portion of the savanna region of central Petén, Guatemala. *Mics. Pub. Mus. Zool. Univ. Mich. 29*.

STUART, L. C., (1941): Studies of Neotropical Colubridae VIII. A revision of the genus Dryadophis Stuart, 1939. *Misc. Publ. Mus. Zool. Univ. Michigan, 49*: 1–106.

STUART, L. C., (1950): A geographic study of the herpetofauna of Alta Verapaz, Guatemala. *Contrib. Lab. Vert. Biol. Univ. Mich. 45*.

STUART, L. C., (1951): The herpetofauna of the Guatemalan Plateau, with special reference to its distribution on the southwestern highlands. *Contrib. Lab. Vert. Biol. Univ. Mich. 49*.

STUART, L. C., (1954): A description of a subhumid corridor across northern Central America, with comments on its herpetofaunal indicators. *Contrib. Lab. Vert. Biol. Univ. Mich. 65*.

STUART, L. C., (1954): Herpetofauna of the southeastern highlands of Guatemala. *Contrib. Lab. Vert. Biol. Univ. Mich. 68*.

STUART, L. C., (1958): A study of the herpetofauna of the Naxactum-Tikal area of northern El Peten, Guatemala. *Contr. Lab. Vert. Biol., Univ. Michigan 75*: 1–30.

STUART, L. C., (1963): A checklist of the herpetofauna of Guatemala. *Misc. Pub. Mus. Zool., Univ. Mich. 122*.

STUART, L. C., (1964): Fauna of Middle America. In: Handbook of Middle American Indians *1*: 316–362. Univ. Texas Press, Austin.

STUART, L. C., (1966): The Environment of the Central American Cold-blooded Vertebrate Fauna. *Copeia 1966* (4): 684–699.

STUTZER, O., (1925): Geographische Beobachtungen an Flüssen und Bächen des mittleren Magdalenen-Tales in Kolumbien. *Peterm. geogr. Mitt.*

SWARTH, H. S., (1929): A new bird family (Geospizidae) from the Galapagos Islands. *Proc. Calif. Acad. Sci. Ser. 4, 18*: 29–43.

SZIDAT, L., (1955): Beiträge zur Kenntnis der Reliktfauna des La Plata-Stromsystems. *Arch. Hydrobiol. 51* (2): 209–260.

TAMSITT, R., (1964): Amphibians, reptiles and mammals of the northern Pacific coastal region of Colombia. *Yearb. Phil. Soc. Amer. 1963*: 353.

TATE, G., (1928): The lost world of Mount Roraima. *Nat. Hist. 28*: 318–328.

TATE, G., (1930): Notes on the Mt. Roraima region. *Geogr. Rev. 20*: 53–68.

TATE, G., (1930): Through Brazil to the summit of Mt. Roraima. *Nat. Geogr. Mag. 58* (5): 584–605.

TATE, G., (1932): Life zones at Mt. Roraima. *Ecology, 3* (3): 235–257.

TATE, G., (1938): Auyan-tepui: Notes on the Phelps Venezuela Expedition. *Geogr. Rev. 28*: 452–474.

TATE, G., (1938): A new 'lost world'. *Nat. Hist. 42* (2): 107–120, 153.

TATE, G. and HITCHCOCK, C. B., (1930): The Cerro Duida Region of Venezuela. *Geogr. Rev. 22*: 31–52.

TAYLOR, E. H., (1951): A brief review of the snakes of Costa Rica. *Univ. Kans. Sci. Bull. 34* (1): 1–188.

TAYLOR, E. H., (1952): The frogs and toads of Costa Rica. *Univ. Kans. Science Bull. 35*: 577–942.

TAYLOR, E. H., (1954): Further studies on the serpents of Costa Rica. *Univ. Kans. Sci. Bull. 36* (11): 673–801.

TAYLOR, E. H., (1955): Additions to the known herpetological fauna of Costa Rica with comments on other species. 2, *Univ. Kans. Sci. Bull. 37*: 499–575.

TAYLOR, E. H., (1956): A review of the lizards of Costa Rica. *Univ. Kans. Sci. Bull. 38*: 1–322.

TAYLOR, E. H., (1958): Additions to the known herpetological fauna of Costa Rica with comments on other species. *Univ. Kans. Sci. Bull. 39* (1): 3–40.

TEAGUE, G. W., (1955): Aves del litoral uruguayo. Observaciones sobre las aves indigenas y migratorias del orden Charadriiformes que frecuentan las costas y esteros del litoral del Uruguay. *Com. Zool. Mus. Hist. Nat. Montevideo 4* (72): 1–55.

TEST, F. H., (1956): Two new Dendrobatid frogs from northern Venezuela. *Occ. Pap. Mus. Zool. Univ. Michigan 577*: 1–9.

THENIUS, E., (1959): Tertiär. 2, Wirbeltierfaunen. Ferdinand Enke Verl., Stuttgart.

THENIUS, E., (1969): Stammesgeschichte der Säugetiere. *Hdb. Zool. 8*: 1–722. Berlin.

THENIUS, E., (1972): Säugetierausbreitung in der Vorzeit. *Umschau 5*: 146–153.

THOMAS, R., (1966): A reassessment of the herpetofauna of Navassa Island. *Jour. Ohio Herp. Soc. 5* (3): 73–89.

THOMAS, R., (1968): The Typhlops biminiensis Group of Antillean Blind Snakes. *Copeia 1968* (4): 713–722.

THOMPSON, W. R., (1963): The Tachinids of Trinidad. *Canad. Entomol. 95* (9): 953–995, (12): 1292–1320.

TIHEN, J. A., (1962): Osteological observations on New World Bufo. *Amer. Midl. Natur. 67*: 157–183.

TIHEN, J. A., (1962): A review of New World fossil Bufonids. *Amer. Midl. Natur. 68* (1): 1–50.

TODD, C. T. and CARRIKER, M. A., (1922): The birds of the Santa Marta region of Colombia: a study in altitudinal distribution. *Ann. Carnegia Mus. 14*: 1–611.

TORDOFF, H. B. and MACDONALD, J. R., (1957): A new bird (family Cracidae) from the early Oligocene of South Dakota. *Auk 74*: 174–184.

TRAYLOR, M. A., (1941): Birds from the Yucatan Peninsula. *Field Mus. Nat. Hist.*, Pub. 493, Zool. Ser., *24*: 195–225.

TREMOLERAS, J., (1920): Lista de las aves uruguayas. *El Hornero 2*: 10–25.

TREMOLERAS, J., (1927): Adiciones y correcciones a la 'Lista de aves uruguayas'. *El Hornero 4*: 16–22.

TRICART, J., SANTOS, M., CARDOSO DA SILVA, T. and DIAS DA SILVA, A., (1958): Estudos de Geografia da Bahia. – Geografia e Planyamento. *Publ. Univ. Bahia 4* (3): 1–243.

TRICART, J., VOGT, H. and GOMES, A., (1960): Note préliminaire sur la morphologie du cordon littoral actuel entre Tramandai et Torres, Rio Grande do Sul, Brésil. *Cah. océanogr. Et. Côtes* Paris *12*: 453–457.

TROLL, C., (1927): Vom Titicacasee zum Pooposee und zum Salar von Coipasa. *Peterm. Geogr. Mitt.*

TROLL, C., (1928): Die zentralen Anden. *Z. Ges. Erdk. Berlin*, Jubil.-Bd. 92–118.

TROLL, C., (1931): Die geographischen Grundlagen der andinen Kulturen und des Incareiches. *Ibero-Amerikanisches Archiv 5*: 258–294.

TROLL, C., (1931): Die Landschaftsgürtel der tropischen Anden. P. M.

TROLL, C., (1950): Savannentypen und das Problem der Primärsavannen. *Proc. Int. Bot. Congr.* Stockholm.

TROLL, C., (1950): Der Vergleich der Tropenvegetation der Alten und Neuen Welt. *Proc. Int. Bot. Congr.* Stockholm.

TROLL, C., (1968): Geo-Ecology of the mountainous regions of the tropical Americas. *Coll. Geogr. 9*: 13–56.

TURK, F. A., (1955): The chilopods of Peru with descriptions of new species and some zoogeographical notes on the Peruvian chilopod fauna. *Proc. Zool. Soc. London 125* (1): 469–504.

UNDERWOOD, G., (1954): The classification and Evolution of Geckos. *Proc. Zool. Soc. London 124*: 469–492.

UNDERWOOD, G., (1954): The distribution of Antillean reptiles. Nat. Hist. Notes, *Nat. Hist. Soc. Jamaica 67*: 121–129.

URETA, E., (1936–37): Lepidopteros de Chile. Rev. Chilena Hist. Nat.

VANDEL, M. A., (1972): De l'utilisation de données biogeographiques dans le reconstitution des anciens visages du globe terrestre. *C. R. Acad. Sc. Paris 274*: 38–41.

VAN DER HAMMEN, TH., (1961): The Quaternary climatic changes of northern South America. *Ann. New York Acad. Sci. 95*: 676–683.

VAN DER HAMMEN, TH., (1966): The Pliocene and Quaternary of the Sabana de Bogotá (the Tilatá- and Sabana formations). *Geol. en Mijnbouw 45*: 102–109.

VAN DER HAMMEN, TH. and GONZALEZ, E., (1960): Upper Pleistocene and Holocene climate and vegetation of the 'Sabana de Bogota' (Colombia, South America). *Leidse Geol. Meded. 25*: 261–315.

VAN DER HAMMEN, T. and GONZALEZ, E., (1964): A pollen diagram from the Quaternary of the Sabana de Bogotá (Colombia) and its significance for the geology of the northern Andes. *Geol. en Mijnbouw 43*: 113–117.

VANZOLINI, P. E., (1963): Problemas faunisticos do Cerrado. Simposio sobre Cerrado. Univ. São Paulo.

VANZOLINI, P., (1968): Geography of the South American Gekkonidae (Sauria). *Arq. Zool. 17* (2): 85–112.

VANZOLINI, P., (1970): Unisexual Cnemidophorus lemniscatus in the Amazonas valley: A preliminary note (Sauria, Teiidae). *Pap. Avul. Zool. 23* (7): 63–68.

VANZOLINI, P. E., (1970): Zoologia sistematica geografia e a origem das especies. Univers. São Paulo, *I.G, 3*: 1–56.

VANZOLINI, P. E. and AB'SABER, A. N., (1968): Divergence Rate in South American Lizards of the Genus Liolaemus (Sauria, Iguanidae). *Pap. Avul. Zool., 21*: 205–208.

VANZOLINI, P. E. and WILLIAMS, E. E., (1970): South American Anoles: The geographic differentiation and evolution of the Anolis chrysolepis species group (Sauria, Iguanidae). *Arq. Zool. 19*: 1–298.

VARESCHI, V., (1955): Monografias Geobotanicas de Venezuela I. Rasgos Geobotanicos sobre el Pico de Naiguata. *Acta Cientifica Venez. 6* (5/6): 1–23.

VAUGHAN, T. W., (1901): Some fossil corals from the elevated reefs of Curaçao, Aruba and Bonaire. Sammlungen des Geologischen Reichs-Museums in Leiden; *Beiträge zur Geologie von Niederländisch West-Indien und angrenzender Gebiete, 2*: 1–92.

VAUGHAN, T. W., (1919): Fossil corals from Central America, Cuba and Porto Rico with an account of the American Tertiary, Pleistocene and recent coral reefs. *Bull. U.S. Nat. Mus. 103*: 189–524.

VELLARD, J., (1941): Serpents venimeux du Venezuela. *Ann. Sci. Nat. Paris, zool. ser. 11* (3): 193–225.

VELLARD, J., (1946): Betracios del Chaco argentino. *Acta Zool. Lilloana 5*: 137–174.

VELLARD, J., (1957): Répartition des Batraciens dans les Andes au sud de l'Equateur. *Travaux Inst. Franc. Etudes. Andines 5*: 141–161.

VERNHOUT, J. H., (1914): The Land- and Freshwater-Molluscs of the Dutch West-Indian-Islands. *Not. Zool. Mus. Leiden 36*: 177–189.

VERTEUIL, J. P. DE, (1968): Notes on the snakes and lizards of Tobago. In: Alford, The island of Tobago, 7th ed.: 101–105. Hampstead, London.

VIAL, J. L., (1968): The Ecology of the Tropical Salamander, Bolitoglossa subpalmata, in Costa Rica. *Rev. Biol. Trop. 15* (1): 13–115.

VIERS, G., (1965): Observations sur la glaciation quaternaire dans les Andes de Mendoza. *Rev. Geogr. Pyr. 36*: 1–116.

VINTON, K. W., (1951): Origin of Life on the Galapagos Islands. *Amer. J. Science 249*: 356–376.

VOOUS, K. H., (1955): Origin of the Avifauna of Aruba, Curaçao and Bonaire. Acta XI Congressus Intern. Ornith., Basel 1954.

VUILLEUMIER, B., (1971): Pleistocene changes in the Fauna and Flora of South America. *Science 173*: 771–780.

VUILLEUMIER, F., (1965): Relationships and Evolution within the Cracidae (Aves, Galliformes). *Mus. Comp. Zool. 134* (1): 1–27.

VUILLEUMIER, F., (1967): Mixed species Flocks in Patagonian Forests, with remarks on inter-species Flock Formation. *The Condor 69* (4): 400–404.

VUILLEUMIER, F., (1967): Phyletic evolution in modern birds of the Patagonian forests. *Nature 215* (5098): 247–248.

VUILLEUMIER, F., (1968): Origin of frogs of Patagonian forests. *Nature 219* (5149): 87–89.

VUILLEUMIER, F., (1969): Biotic Diversity and Environmental Stability. *Science 166*: 210–211.

VUILLEUMIER, F., (1972): Bird species Diversity in Patagonia (Temperate South America). *Americ. Natural. 106*: 266–271.

WAIBEL, L., (1921): Urwald, Veld, Wüste. Breslau.

WAIBEL, L., (1948): Vegetation and land use in the Planalto Central of Brazil. *Geogr. Rev.* 38.

WAGENER, S., (1961): Monographie der ostasiatischen Formen der Gattung Melanargia Meigen (Lep., Sat.). Stuttgart.

WAGNER, H. O., (1961): Die Nagetiere einer Gebirgsabdachung in Südmexiko und ihre Beziehungen zur Umwelt. *Zool. Jb. Syst.* Bd. 89: 177–242.

WAGNER, M., (1870): Naturwissenschaftliche Reisen im tropischen Amerika. Stuttgart.

WAGNER, M., (1889): Die Entstehung der Arten durch räumliche Sonderung. Basel.

WALKER, CH. F. and TEST, F. H., (1955): New Venezuelan frogs of the genus Eleutherodactylus. *Occ. Pap. Mus. Zool. Univ. Michigan 561*: 1–10.

WALLACE, A. R., (1876): Geographical distribution of animals. London.

WALLACE, A. R., (1880): Island life. London.

WALTER, H., (1954): Die Verbuschung, eine Erscheinung der subtropischen Savannenlandschaft, und ihre ökologischen Ursachen. *Vegetatio 6*, Den Haag.

WALTER, H., (1962): Die Vegetation der Erde in öko-physiologischer Betrachtung. Stuttgart.

WALTER, H., (1966): Das Pampaproblem und seine Lösung. *Ber. Dtsch. Bot. Ges. 79* (8): 377–384.

WARNECKE, G., (1961): Über atlanto-mediterrane Großschmetterlings-Arten in Norddeutschland, insbesondere im Niederelbgebiet und in Schleswig-Holstein. *Verh. Ver. naturw. Heimatf.. Hamburg 35*.

WEBB, R. G. and BAKER, R. H., (1962): Terrestrial Vertebrates of the Pueblo Nuevo Area of southwestern Durango, Mexico. *Amer. Midl. Natur. 68*: 325–333.

WEBER, H., (1958): Die Paramos von Costa Rica und ihre pflanzengeographische Verkettung mit den Hochanden Südamerikas. Akad. Wiss. u. Literatur, *Abhdl. Math.-Nat. Kl. 3*: 121–195.

WEBER, H., (1959): Los Paramos de Costa Rica y su concatenación fitogeografica con los Andes sudamericanos. San José, Costa Rica, Minist. Obr. Publ.

WEBER, H., (1969): Zur natürlichen Vegetationsgliederung von Südamerika. In: Biogeography and Ecology in South America 2: 475–518.

WEBERBAUER, A., (1911): Die Pflanzenwelt der peruanischen Anden. Leipzig.

WEBERBAUER, A., (1945): El mundo vegetal de los Andes Peruanos (Estudio fitogeografico). Ministerio de Agricultura, Lima.

WEIGEL, J., (1969): Systematische Übersicht über die Insektenfresser und Nager Nepals nebst Bemerkungen zur Tiergeographie. *Khumbu Himal 3*: 149–196.

WEISCHET, W., (1964): Geomorfologia glacial de la region de los lagos. Univ. Chile, Santiago de Chile.

WELLMAN, J., (1963): A revision of snakes of the genus Conophis (Family Colubridae), from Middle America. Univ. Kansas Publ. *Mus. Nat. Hist. 15*: 251–295.

WERMUTH, H., (1965): Gekkonidae, Pygopodidae, Xantusiidae. In: *Das Tierreich, 80*: 1–246, Berlin.

WERNER, D. J., (1972): Campo Arenal (NW-Argentinien). Eine landschaftsökologische Detailstudie. Biogeographica *1*: 75–86.

WERNER, F., (1896): Beiträge zur Kenntnis der Reptilien und Batrachier von Zentralamerika und Chile. *Verhdl. Zool. Bot. Ges. Wien 46*.

WERNER, F., (1897): Die Reptilien und Batrachier der Sammlung Plate (Fauna Chilensis). *Zool. Jb. Suppl. 4*: 244–278.

WERNER, F., (1916): Bemerkungen über einige niedere Wirbeltiere der Anden von Kolumbien mit Beschreibungen neuer Arten. *Zool. Anz. 47* (11): 301–304.

WEST, R. C., (1957): The Pacific lowlands of Colombia. Louisiana State Univ. Studies, social sci. ser. *8*: 1–277.

WESTERMANN, J. H., (1932): The geology of Aruba. Geogr. Geol. Meded. Utrecht Phys. Geol. *7*.

WETMORE, A., (1923): Avian fossils from the Miocene and Pliocene of Nebraska. *Bull. Amer. Mus. Nat. Hist. 48*: 483–507.

WETMORE, A., (1926): Observations on the birds of Argentina, Paraguay, Uruguay and Chile. *Bull. U.S. Nat. Mus. 133*: 1–448.

WETMORE, A., (1933): A fossil gallinaceous bird from the Lower Miocene of Nebraska. *Condor 35*: 64–65.

WETMORE, A., (1941): New forms of birds from Mexico and Colombia. *Proc. Biol. Soc. Washington 54*: 203–210.

WETMORE, A., (1946): New birds from Colombia. *Smithsonian Misc. Coll. 106*: 1–14.

WETMORE, A., (1951): Recent additions to our knowledge of prehistoric birds, 1933–1949. Proc. X. Intern. Ornith. Congr. Uppsala *1950*: 51–74.

WETMORE, A., (1953): Further additions to the birds of Panama and Colombia. *Smithsonian Misc. Coll. 122*: 1–12.

WETMORE, A., (1956): A fossil guan from the Oligocene of south Dakota. *Condor 58*: 234.

WETMORE, A. and PHELPS, W. H., (1956): Further additions to the list of birds of Venezuela. *Proc. Biol. Soc. Washington 69*: 1–12.

WEYL, R., (1955): Vestigos de una glaciacion del Pleistocene en la Cordillera de Talamanca, Costa Rica, A. C. Inst. Geog. Costa Rica, 9–32.

WEYL, R., (1956): Spuren eiszeitlicher Vergletscherung in der Cordillera de Talamanca Costa Ricas (Mittelamerika). *Neues Jb. Geol. Paläont., Abhdl. 102*: 283–294.

WEYL, R., (1956): Geologische Wanderungen durch Costa Rica. Natur und Volk, *86*.

WEYL, R., (1956): Eiszeitliche Gletscherspuren in Costa Rica (Mittelamerika). *Z. Gletscherkd. u. Glazialgeol. 3*: 317–325.

WHEELER, W. M., (1919): The ants of the Galapagos Islands. The ants of Cocos Island. *Calif. Acad. Sci. Proc. 2*: 259–308.

WHITE, S. E., (1956): Probable substages of glaciation on Ixtaccihuatl, Mexico. *J. Geol. 64*: 289–295.

WHITELY, H., (1884): Explorations in the neighbourhood of Mounts Roraima and Kukenam, in British-Guiana. *Proc. Roy. Geogr. Soc. 6* (8): 452–463.

WILHELMY, H., (1952): Die eiszeitliche und nacheiszeitliche Verschiebung der Klima- und Vegetationszonen in Südamerika. Dt. Geographentag Frankfurt, Remagen.

WILHELMY, H., (1954): Die klimamorphologische und pflanzengeographische Entwicklung des Trockengebietes am Nordrand Südamerikas seit dem Pleistozän. *Die Erde 6*: 244–273.

WILHELMY. H., (1954): Paraguay, das Land und seine Menschen. *Mitt. Inst. Auslandsbez. 5*: 275–281. Stuttgart.

WILHELMY, H., (1957): Eiszeit und Eiszeitklima in den feuchttropischen Anden. Geomorphologische Studien (Machatschek-Festschrift), Ergänzungs-Heft *262* zu *Peterm. Geogr. Mitt.* 281–310.

WILLEY, G. G., (1946): Archeology of the Greater Pampa. Hdb. South Americ. Indians, Bur. *Am Ethn. Bull. 193*: 25–46, Washington.

WILLIAMS, E., (1950): Testudo cubensis and the evolution of western Hemisphere tortoises. *Bull. Amer. Mus. Nat. Hist. 95* (1): 1–37.

WILLIAMS, E., (1960): Two species of tortoises in northern South America. *Breviora 120*: 1–13.

WILLIAMS, E. and KOOPMANN, K. F., (1951): A new fossil rodent from Puerto Rico. *Amer. Mus. Novitates 1515*: 1–9.

WILLIAMS, E. and KOOPMANN, K. F., (1952): West Indian fossil monkeys. *Amer. Mus. Novitates 1546*: 1–16.

243

WILLIAMS, E. E. and VANZOLINI, P. E., (1966): Studies on South American Anoles, Anolis transversalis A. DUMERIL. *Pap. Avulsos 19*: 197–204.
WILLIAMS, F. X., (1911): The butterflies and hawk-moths of the Galapagos Islands. *Calif. Acad. Sci. Proc. 1*: 289–322.
WILLIAMS, F. X., (1926): The bees and aculeate wasps of the Galapagos Islands. *Calif. Acad. Sci. Proc. 2* (2): 347–357.
WILSON, E. O. and SIMBERLOFF, D. S., (1969): Experimental zoogeography of Islands: defaunation and monitoring techniques. *Ecology 50*: 267–278.
WINKLER, O., (1926): Niederländisch-Westindien. *Mitt. Ges. f. Erdk.*, 87–137.
WOLDSTEDT, P., (1958): Das Eiszeitalter. Stuttgart.
WOLFFHÜGEL, K., (1929): La exploracion de la alta Cordillera de Mendoza. Buenos Aires.
WOODRING, W. P., (1966): The Panama land bridge as a sea barrier. *Proc. Am. Philos. Soc. 110* (6): 425–433.
WRIGHT, J. W. and LOWE, CH. H., (1968): Weeds, Polyploids, Parthenogenesis, and the Geographical and Ecological Distribution of All-Female Species of Cnemidophorus. *Copeia 1968* (1): 128–138.
WYMSTRA, T. A. and HAMMEN, T. VAN DER, (1967): South American savannas: Comparative studies, Colombia and Guayana. McGill University, Montreal.
ZERRIES, O., (1968): The South American Indians and their Culture. In: Biogeography and Ecology in South America *1*: 329–388.
ZIMMER, J., (1931–1955): Studies of Peruvian birds. *Amer. Mus. Novit. 930.*
ZIMMER, J. and PHELPS, W., (1944): New species and subspecies of birds from Venezuela. I. *Amer. Mus. Novit. 1270*: 1–16.
ZIMMER, J. and PHELPS, W., (1946): Twenty-three new subspecies of birds from Venezuela and Brazil. *Amer. Mus. Novit. 1312*: 1–23.
ZIMMER, J. and PHELPS, W., (1948): Three new subspecies of birds from Venezuela. *Amer. Mus. Novit. 1373*: 1–7.
ZIMMER, J. and PHELPS, W., (1949): Four new subspecies of birds from Venezuela. *Amer. Mus. Novit. 1395*: 1–9.
ZIMMER, J. and PHELPS, W., (1952): New birds from Venezuela. *Amer. Mus. Novit. 1544*: 1–7.
ZWEIFEL, R. G., (1965): Distribution and mating calls of the Panamanian toads. Bufo coccifer and B. granulosus. *Copeia 1965*: 108–110.

Address: Prof. Dr. PAUL MÜLLER, Department of Biogeography, Universität des Saarlandes, 66 Saarbrücken.